Samuel Marshall Colcord

Colcord's System of Preserving Green Forage Without Heat or Fermentation

Samuel Marshall Colcord

Colcord's System of Preserving Green Forage Without Heat or Fermentation

ISBN/EAN: 9783744692502

Printed in Europe, USA, Canada, Australia, Japan

Cover: Foto ©berggeist007 / pixelio.de

More available books at **www.hansebooks.com**

COLCORD'S SYSTEM

OF

Preserving Green Forage

WITHOUT

HEAT OR FERMENTATION

BY THE USE OF

THE SILO GOVERNOR

BY

SAMUEL M. COLCORD

DOVER MASS.

CHICAGO ILL.
HOWARD & WILSON PUBLISHING CO.
1889

DEDICATION.

This volume is dedicated to the Howard & Wilson Publishing Company, Chicago, from my warm regard and gratitude to them as the first persons to perceive and candidly acknowledge, through the columns of their valuable journal, the great importance and merit of my system of perfectly preserving green forage.

The Author.

CONTENTS.

	PAGE
PREFACE,	7
THE SYSTEM AND DEVICE,	13
OF SILOS,	15
FILLING AND EMPTYING THE SILO,	19
THE FEED-BOX,	21
CUT OF THE FEED-BOX,	24
DESCRIPTION OF THE FEED-BOX,	25
WEIGHTING THE SILO,	26
THE CROP TO PRESERVE,	30
MANAGEMENT OF THE CROP,	31
FERMENTATION IN SILOS,	34
THE SILO GOVERNOR,	37
CUT OF SILO GOVERNOR,	41
DESCRIPTION OF THE SILO GOVERNOR,	42
A HALF EMPTY SILO,	48
SILAGE *versus* DRY FODDER. By *Professor Arnold*,	48
EXPERIMENTS WITH ENSILAGE. By *Dr. E. L. Sturtevant, Director of the State Experimental Station at Geneva, N.Y.*,	50
THE OPINIONS OF EMINENT AGRICULTURISTS,	51
SWEET FORAGE IN WINTER. From the *Farm, Field, and Stockman*,	53
MY EXPERIMENT SILO,	55
BUILDING SILOS,	64
DIRECTIONS FOR PUTTING IN AND REMOVING THE SILO GOVERNOR,	69
CUT OF THE TABLE-TOP CORN-CART,	74
DESCRIPTION OF THE TABLE-TOP CORN-CART,	75
FAULTY SILOS AND FAULTY MANIPULATIONS,	76

Contents

	PAGE
REMEDY FOR FAULTY SILOS,	80
ENSILAGE AND ITS IMPORTANCE. From the *Dairy World*,	82
ELIMINATING THE AIR,	85
PRESERVING GREEN FOOD. Something New and Important in Live Stock Economy. From the *Indiana Farmer*, 1887,	85
THE COLCORD ENSILAGE EXPERIMENTS. From the *Farm, Field, and Stockman*, 1888,	87
WHAT MY NEIGHBORS SAY,	90
WHAT THE BUTCHER SAYS,	97
PRIVATE CORRESPONDENCE,	99
WHAT LARGE DAIRYMEN SAY,	108
PRESERVED GREEN FORAGE FED TO YOUNG CALVES,	110
THE SILO GOVERNOR. From the *Massachusetts Ploughman*,	112
ABOUT FERTILIZERS,	115
SILOS AND ENSILAGE. From the *New England Farmer*,	122
SWEET ENSILAGE. From the *New England Farmer*,	127
ENSILAGE A PROMOTER OF DIGESTION AND ASSIMILATION,	130
COLCORD'S PRESERVED GREEN FORAGE. From the *Farm, Field, and Stockman*, 1889,	132
EXPERIMENTS WITH MILK AND CREAM. From the *Farm, Field, and Stockman*, 1888,	138
PATENT SILAGE. From the *Rural New Yorker*,	141
A SUGGESTION FOR THE EXPERIMENT STATIONS. From the *New England Farmer* and *Rural New Yorker*, 1889,	142
THE PRESERVATION OF ENSILAGE. From the *Report of the State Board of Agriculture of Pennsylvania*, in 1888. By S. M. Colcord,	143
PROGRESS MADE IN PRESERVING GREEN FORAGE IN SILOS,	157

PREFACE.

THIS little treatise is designed to give full information and explanation of Colcord's method and device for Preserving Green Forage; and I have endeavored to write in plain, direct language, so that all persons interested in the subject may be able to readily understand and work by this system without difficulty.

The treatise is arranged to give our present knowledge on its first pages; then the means pursued by which the knowledge was obtained; the proofs, tests, experiments, progressive experience, theories, and certificates are given later on. This arrangement necessarily causes considerable repetition, as well as apparent contradiction; but, as much of the matter embraced in this treatise has been published from time to time, as facts have been developed, I prefer to reprint some of the original articles, with the comments made by the press at the time of their first appearance, remarking that any investigator will be a lucky man who will investigate as many years as I have and not find occasion to change his theories and opinions quite as often as I have done.

It should be remembered that it consumes one whole year's time to make each experiment, or each class of experiments, and that it is necessary to verify the work in our own silo by the labors of other men with other silos. My silo, manipulations, and results are always open to the investigations of others.

Everybody ought to know how utterly impossible it is for any one man to make successful experiments in opposite directions at the same time, with opposite systems, theories, and modes of operation: one with heat, another without heat; one cutting forage very fine, another packing it in whole; one weighting with portable weights of bags, boxes, or barrels, another pressing with screws; one cutting down vertically, another forking off from the whole top; one making an ensilage more or less repulsively odorous, another pressing out juice in quantity, bringing it throughout to the top of the silo, removing the air and free gases, and producing a wholesome, nutritious food, without waste or odor.

I take great pleasure in thanking the press

for what they have done to bring my system and device to the notice of agriculturists; especially the *Farm, Field, and Stockman*, the *Dairy World*, the *Indiana Farmer*, the *New England Farmer*, the *Rural New Yorker*, the *State Board of Agriculture of Pennsylvania*, in their Report of 1888, the *Scientific American*, and several papers in foreign languages that have volunteered to publish and illustrate this invention as a public benefit; also, those papers that have advertised my system and governor, without admitting me to their editorial columns, although they state that they advertise nothing that they cannot recommend.

All the editorial and other matter herein presented as copied from those papers has been published for the benefit of the readers of the above-named papers without expense to me, excepting the use of my engravings for illustration.

I also insert, with these previously published accounts, the remarks of the *former* editor of the *Massachusetts Ploughman* in relation to the meeting and investigations of the farmers as to the merits of the silo governor, at which meeting the following closing remarks were made by Mr. Ware:—

" The chairman of the meeting, at its opening, stated that what was wanted was real experience instead of theory. It is but fair to state that Mr. Colcord has confined his remarks to practical experience and proofs, with corroborating testimony about the governor."

It may seem contradictory, or merely a matter of opinion, that capillary attraction should be stated as the cause of bringing the juice from the bottom to the top of the silo, and holding it there, when it is also stated that carbonic acid had taken the place of the air in the silo, and that under pressure the carbonic acid was absorbed by the juice, causing a partial vacuum, which is the cause of the rising of the juice to the top of the silo; but both these statements are true, either separately or combined.

When I had 30 inches of juice at the bottom of the silo, and was pressing heavily, it happened that the juice began to disappear very rapidly. I did not believe that capillary attraction could be the cause of it, and supposed that the pressure had burst a hole in the bottom of the silo; but, when the silo was empty, I could find no leak. I allowed the water from the aqueduct to run into the silo for an entire day, kept the water in it for a month, and found the silo perfectly tight. I then remembered that, before the silo was opened, carbonic acid had disappeared, and that acetic acid had remained in, or continued to come to the top of the perpendicular pipe above the silo. This was proof positive of what had taken place.

We wash down the walls of the silo with water, using for that purpose a long-handled whitewash brush. This water is all drawn off through the governor drip pipe, and this is what is meant by using the governor to draw off water. We never put water into the forage; the corn contains more than we have occasion to use, and this year (1889) we have been feeding as high as 100 pounds daily of juice, but reduced the rations to 50 pounds until the excess of juice was used up.

The difference in the corn crop between 1887 and 1888 was very marked, and accounts for much of the difference in our results. The last year's crop was badly frost-bitten, quite immature, gathered and packed very wet, consequently it required much less pressure and gave a much larger proportion of free juice; but the screws and the governor entirely controlled these inequalities, and, so far as the preservation was concerned, the result was perfectly satisfactory.

I will here express my thanks to those gentlemen who have encouraged me to prosecute my investigations, also to those who have given the governor a fair trial and are now giving me their aid and encouragement. I wish also to draw the particular attention of the reader to the certificate of the marketman who butchers my cows, fatted upon this forage and wheat bran, (no corn meal being used); also to the *New York Experiment Station*, in making the experiment of mixing acetic acid with green corn forage. It is valuable as showing the good effects of a limited quantity of acetic acid, and coincides with my experience. But I don't think it safe to recommend such large addition of acetic acid, as a steady diet, to ensilage as commonly prepared; the cows, as a general thing, get too much of it.

Seeing little or nothing more to be accomplished, I now offer this as my Perfected System of making *Preserved Green Forage without Heat or Fermention.*

SAMUEL M. COLCORD,
Dover, Mass.

THE SYSTEM AND DEVICE.

THE name of "ensilage" has been applied to all kinds of green forage crops that have passed through silos. It was first introduced into this country about twelve years ago, through the publications of Monsieur Auguste Goffart, of Sologne, France. The art of preserving green forage without desiccation has often been attempted, and has been traced back to remote antiquity; but to M. Goffart belong the invention and introduction of ensilage through silos, and to him we accord the honor. Any person who has made a study of his theory and practice, and who has studied the art as practised in this country up to the present time, will agree with me that the closer one follows M. Goffart's system, and the less he follows the professed improvements on his system, as practically illustrated in this country, the better will be his ensilage. And I feel warranted in making the assertion that

M. Goffart was producing better ensilage in France, twelve years ago, than is being made by a vast majority of his followers in this country to-day, notwithstanding all their experiments and attempted improvements upon his system.

M. Goffart, in his writings, makes this statement: —

"The end to be attained is *to prevent* ALL KINDS OF FERMENTATION, before and after ensilage. *Fermentation preserves nothing;* on the contrary, *it is always a preliminary step towards a decomposition more or less putrid,* towards a REAL DESTRUCTION."

M. Goffart always worked to attain the end above expressed, as nearly as possible; and, although he claimed a perfect system and was very particular in his manipulations, his writings show that he never fully realized the end he sought, but always speaks of his ensilage as heating up when exposed to the air, taking on the alcoholic fermentation, then running into the acetic, and finally passing into the lactic and other putrid fermentations. This would not, could not, have been the case if he had never had heat and fermentation in his silo.

My experience is that, when there has been

no fermentation in the silo, the forage does not heat up and pass through the alcoholic and acetic fermentation to lactic and putrid, but sometimes takes on a mouldy condition, which develops black rot and causes destruction in that way. This may be called fermentation, but it is not a true fermentation.

That I have succeeded, after years of study and costly experimenting, in perfectly removing air from the silo, preventing heat and fermentation, and Preserving Green Forage Corn in perfection, will be demonstrated in the following pages.

OF SILOS.

A really good silo must be tight and strong and impervious to air and water. It should have a good foundation, perfectly drained and perfectly perpendicular, smooth, level-faced walls. If these conditions are fulfilled, it is not very material of what they are made; but, when made of masonry, all forms of lime must be excluded, as acetic acid dissolves the lime. Cement must be used instead of lime mortar.

Good silos are somewhat expensive; but true economy points in the direction of durability, convenience and assured success. Their attachment to the barn, for convenience, should be provided for; and also the means for pressing the forage, which is a very important item in economy, time, and convenience. Pressing by jack-screws, if properly arranged, is the most simple, convenient, economical, and successful. It is accomplished by putting iron rods, 1¼ inches in diameter, in the centre of the side walls, from the foundation up to from 4 to 6 feet above the top of the silo, said rods being made with broad flanges on their lower ends, and long screws on their upper ends provided to receive double nuts and large washers, these rods to be placed in rows commencing and ending 4 feet from each end of the silo, and not more than 8 feet apart, arranged on both sides alike. The opposite rods should be tied together across the top, with 8 x 8 timbers provided with holes, so that they may slip loosely upon the rods. The cap of the wall should be 6 x 8 timber, set back 2 inches from the inside face of the walls, to receive the 2-inch plank placed around the top of the silo for the purpose of building a light annex, or head-room, from 3 to 6 feet higher than the silo proper, said annex to be filled with forage and answering the purpose of so much solid wall. The forage placed in the annex, when pressed, will all come down inside the solid walls. Under strong pressure, these walls are held up very firmly by the iron rods, the timber across the top acting as a spring upon the forage. In this way, we feel sure of the strength of our walls, and we can get all the pressure we want. The governors convey the abundance of juice to all parts of the silo evenly. 2 x 8 plank studding, to support the roof of the silo, should be placed upon the cap, so as to support the planks placed round the top, and bring them just level with the silo wall. In this way, all time, trouble, and expense of weighting are avoided. In weighting, when the governor is used, it requires about 100 pounds to each square foot of surface, which is equal to 200 or 300 pounds where there is no governor.

In the very centre of the bottom of the silo is placed one end of the drip pipe (seen in cuts, Figs. 1, 2, letter k, page 41, to come flush with the surface of the bottom. This pipe should be about 3 inches long, made of 1½-inch pipe, screwed into an elbow at the bottom of the silo, and from this elbow should run a horizontal pipe declining 6 inches to any convenient place outside and from underneath the wall. This pipe should end in a ⊥ to turn up, with a stop-cock in the end of the ⊥ to draw off the juice, and an upright pipe from the ⊥ to come up outside the silo (see cut, Figs. 1, 2, letter l, p. 41), for the escape of the air and gases. The drip pipe forms a part of the silo governor, and the stop-cock comes over a little well for convenience in drawing off juice.

It is usual to have 8 or 9 feet head-room above the silo wall, for convenience, and to fill the silo above the wall up into the annex, for economy. This head-room is also useful for storing the plank covering for the silo; the planks that go around the top of the silo to build the annex; the 6 x 8 timber that runs the length of the silo across the cover upon which the jack-screws are placed, and any other timber or article of use. The jack-screws and blocking are placed upon the cap between the studding, so that all the timber and tools are kept at the top where they are wanted and do not have to be lowered or hoisted. Each plank, as it is removed, is placed in the head-room; and, when the forage is all fed out, everything is in place for the next season, and the silo is entirely empty. Viewed from the bottom, it appears impossible for any one to get at the top for theft or disarrangement.

FILLING AND EMPTYING THE SILO.

We do not care to tread down the corn as we fill the silo, but only keep it level and walk over it for the purpose of finding soft spots, which we fill up level. When we get the silo full, we tread it hard and level, rounding it up over the top, even above the cross timbers, and allow it to remain until the next morning, when we level it and tread again; then put on cover, and then the 6 x 8 cross timbers 2 feet from the side walls, placing the jack-screws between the timbers. The governor being in place, we drop a thermometer, appended to a string, into the governor, to the centre of the silo. We also put a stick in the upright pipe of the bottom governor, said stick being long enough to touch the bottom of the pipe, for the purpose of measuring the juice. We take this measure daily; also take the temperature daily, and press as often as required. All the corn is cut to half-inch pieces, and is therefore a homogeneous mass. When we press it, we mark a long stick to feet and inches, setting it opposite to one of the screws. We then turn down that screw to the mark we wish them all to go to, taking the stick to the next screw, and so on, pressing all alike, measuring from the top of the silo around the wall. By so doing, we press all the contents of the silo to a uniform density, the forage slipping in the silo, and finding its level and density like mortar in a bucket.

Now, we are supposed to have a silo full of this preserved forage, 20 feet deep, consisting of from 1 to 400 tons. We have a door 4½ x 6 feet, in one end of the silo, said door opening upon the barn floor, the door-sill being 10 feet from the bottom of the silo. We now roll the door to one side, and find some boards tacked onto the edges of 2 x 8 plank, fastened with 4-inch lag-screws to the 3 x 12 inch door jams, on each side, leaving the 2-inch matched plank flush with the inside of the silo. We remove the boards, then the wet sand between the boards and plank, then the side plank, fastened with the lag-screws, then the inside plank, the removal of which presents a solid wall of green forage, with every particle of it in perfect preservation and ready to feed out, the doorway having been secured air-tight. We cut this preserved forage down, vertically and evenly, with a sharp lightning-hay-knife, leaving a solid, smooth face, which prevents the air from getting into the forage. We then discover that more than half the feed has to be elevated from 1 to 10 feet to the barn floor.

We find that we can easily elevate or depress the cross timbers upon the iron rods any required distance, blocking the ends between the cap and the washers. We then run a line of 6 x 6 timber the whole length of the silo upon the 8 x 8 cross timbers in the centre, securing them with 10-inch lag-screws. We fasten the hangers of the track to the bottom of the 6 x 6 timber in the centre with 4-inch lag-screws. We then place the $3/8$ x 3 inch iron in the slots at the bottom of the hangers, fastening them by turning up the set-screws, and the track is complete. This track is furnished with a double-roller troll, and is quickly put up by any one of ordinary capacity, and is easily changed or removed with very little time or trouble. The double-roller troll, with the track and hangers, is made by R. J. Davies, Creek Square, Boston, and costs from $15 to $20.

THE FEED-BOX.

A feed-box, made of stock $1\frac{1}{2}$ inches thick, 2 feet deep, 4 feet wide outside at the top, and 3 1-3 feet wide outside at the bottom, with perpendicular ends grooved into sides 5 feet long, $1\frac{1}{2}$ inches from the ends, with the bottom projecting 1½ inches all around, securely nailed on, will hold enough to feed 25 head of cattle. Being in constant use, it should be well ironed all around the top and down the ends, at the sides, and securely fastened together by ⅜ iron rods, with nuts upon each end, three of them across each end, going through irons on the outside. This box is suspended (see cut, p. 24) on a ½ ton compensated chain hoist by a chain on each side of the box, with a ring in the centre, the ends of the chains going through the eye of an iron at the top of the box, the other end of the iron being fastened to the ⅜ rod running across the outside ends of the box in the centre; also, by bolts near the chain, so that the chains will be in line from the ring in the centre to the ⅜ rods supporting the box from the four corners at the centre, the rings in the centre being hooked to the chain hoist.

An iron axle, with wheels 6½ inches in diameter, is securely fastened across the bottom of the box, 22 inches from one end; the wheels have 1½ inch tread, and run close to the box, and have a wooden shield to protect them from the chain. The other end of the box should be run upon a good strong castor in the centre. This will enable the box to turn and run in any direction from the silo to every cow.

With this device, one man can feed all the cattle with less labor than in any other way; in fact, it seems more like sport than work. The box runs the whole length of the silo, and remains just where we want it, full or empty, at any elevation we may be removing the forage from the face of the cut. We do not have to lug any of it. Thus arranged, the device seems to be indestructible, and time, trouble, and labor reduced to the minimum. There is no waste, litter, or odor about the barn or silo. The box rolls out of the silo upon a Fairbanks scale, every ration is weighed, and it is all eaten up clean.

The cattle require only about one-half the usual quantity of water: ours drink no cold water, and the results are shown at the milk-pail, the scales, and the manure pit. In fact, so quickly and quietly is this shown in practical operation that it takes less time to show it than it does to tell and explain it.

Colcord's System of

THE FEED-BOX.

DESCRIPTION OF THE FEED-BOX.

The cut represents a feed-box capable of holding the rations for 25 head of cattle, showing its construction, also the compensated chain hoist and the double-roller troll which runs upon the track over the silo. When the box is lowered upon the barn floor, the chains are unhooked from the hoist, and the rings, by which it is shown as suspended, are placed in the hooks upon each side of the box, leaving the top of the box perfectly free and without any obstruction, the wheels at the bottom allowing the box to turn and run in any direction. In practical use, it works perfectly, and is found to be the most convenient and expeditious way of conveying the forage from the silo to the cattle.

The feed-box is hoisted and lowered by an endless chain. Said chain is shown in the cut as hanging loosely against the sides of the box; but quite a large portion of its length is omitted in the cut, for the purpose of saving the room it would occupy on the page of the book. The two broken ends from which the omitted portion was severed are shown in the cut as hanging a little below the bottom of the box.

WEIGHTING THE SILO.

Weighting has always been the great objection to silos: how to put on and take off 20 to 40 tons of weight, when time is precious, and to do it cleanly and neatly, keeping dirt, stones, chips, etc., out of the forage, and not making a litter about the barn, to say nothing of the continual expense, especially when the weighting has to be hoisted and lowered. Boxes and barrels are constantly coming to pieces, and not convenient to handle. Of course, the weight upon the forage is what we must have, and the cheapest way to get it there is by some thought to be the best; but, in taking the weight off, it is very important to keep it upon every plank that you do not take off, to enable you to make the vertical cut on the forage when removing it to feed out. I will here suggest the best method of weighting. Take a piece of board, 1 inch thick, 12 inches long, and 16 inches wide, for the bottom of a box, 2 pieces 18 inches long and 24 inches wide for the sides, and 2 pieces 25 inches long and 10 inches wide for the ends. Place the sides *on* the upper surface of the bottom, and the ends on the vertical ends of the bottom; nail them firmly together, and you will have a box in which none of the nails will be driven into the wood parallel with the grain, and each attachment will act as a cleat across the boards, to prevent their warping or splitting. These boxes are intended to be placed close together upon the 2 x 12 inch plank covering the silo, said boxes having no cleats on the outside. Put cleats 2½ inches wide, bevelled at the top, across the ends on the inside of the box. Two men can use these cleats as handles, and also to hoist with, by using an ⋀ shaped iron having an eye in its top, and turned out one inch each side at the bottom, to go under the cleats. They may be filled with sand or loam for bedding and to absorb liquid manure, or may be used for gravel for weighting only. For strength, durability, and convenience, they are unrivalled. The word "long," as used above, means measuring *with* the grain, and the word "wide" means measuring crosswise of the grain. As it is sometimes very difficult to find boards 24 inches wide crosswise of the grain, 4 pieces 18 inches long and 12 inches wide will serve to make the sides, instead of the 18 x 24 inch stuff.

Two hundred and fifty boxes, made in the above manner, at 25 cents each, would cost $62.50. Flour barrels would cost about half that price, and iron rods and jack-screws would cost about $62.50. The boxes would last 31 years, or $2.00 a year; the barrels, 3 years, or $10.00 a year; the screws and rods, 70 years, or, say, $1.00 a year. But the time, accuracy, and convenience of the screws would more than double the economy of the boxes or barrels, and should be reckoned at only 50 cents a year. I esteem the value of the rods, for strength and security, fully equal to their value in pressing the forage; and, if I were building a cheap wooden silo, I should put them in, first building a good 18-inch cement wall, 3 feet high, upon a good foundation, well drained, setting my wood silo upon it, putting a timber between each rod and the inside double boarding, with 2 x 12 inch studding, filled around the bottom with cement and gravel, and between the out and inside boarding with sawdust to keep out the frost. I would also spike 2-inch plank firmly around the outside, at the top, middle, and bottom, because you want to be sure of your 8 or 10 months' food for your cattle, and silos are so difficult and expensive to repair, if the walls give way. Built in this way, wooden silos are easily converted into cement ones, which are sure to be wanted in the future, and are always permanent and require no repairs.

There are other methods of getting pressure, — with levers, also with water; but these are no cheaper and not so convenient, because with jack-screws, costing from $2 to $4 each, you can remove the screws and blocking at pleasure, and set them back as you cut down, keeping the pressure on, which is a great advantage. But with water, even if you have an aqueduct to run the water in and out the barrels with a hose, the water may freeze; and the barrels are always in the way and cost more than to press with screws and rods.

It does not work well to have a double cover running lengthwise and crosswise the silo, for you cannot remove part of it at a time. The best cover is 2-inch plank, laid directly upon the forage, with 6 x 6 timber laid across lengthwise, about 3 feet from the side walls, to keep the plank level. Uncover no faster than you cut down. I have found by repeated examinations that, when you uncover the whole top to feed out, by forking it off, the top, owing to exposure to the air, is about twice as sour as it is about 2 feet below where you fork from, so that the stock get twice as much acid as there is any occasion for, and often more than is healthy.

THE CROP TO PRESERVE.

Almost any kind of green forage can be preserved in silos. The general conditions to be observed, in putting it in the silo, are to have it a homogeneous mass when pressed. For this reason it should be cut fine, especially when the stalks are coarse and hard like corn. It is much better to have but one kind of fodder in the silo at the same time, for the reason that the softest kind, if more than one variety is used, will pack quicker, and hard enough to prevent the escape of air and gas. The air and gas will collect in spots, and set up heat and fermentation; but, if the mass is in uniform condition, evenly spread and pressed, the air may all be removed from it without difficulty, which will insure good preserved forage.

There may be cases where there is a heavy crop of coarse marsh grass, fresh or salt, which does not require cultivation, and is convenient to the silo, that would pay to cut up and preserve, in which case it would greatly enhance its value for feeding.

There are sections at the South where some heavy crops grow without cultivation that would make good feed for cattle, and would be greatly increased in feeding value by preservation in the silo, but which would be comparatively valueless preserved by desiccation. A case in point is the Roman wormwood, the common ragweed of the North, which I have known to be ensiloed, and is said to have made a very palatable food. But to raise a crop of anything else for the silo, on land upon which Indian corn will grow, seems to be a waste of time and money. It is better to plough under the crop of weeds before they go to seed, and plant a crop of mammoth ensilage corn. When you do this, you are feeding the land as well as the cattle, at the same time, much more economically.

MANAGEMENT OF THE CROP.

Indian corn, above all other plants, is the crop for the silo, because it is the best food, is greatly increased in feeding value by soaking in its own juice in the silo under pressure, is a great appetizer in this form, is more assimilable as food, and the plant, or corn, in the milk does not have the injurious effect of cornmeal. From 20 to 40 tons can be raised to the acre of land, 3 tons of it being equal in feeding value to 1 ton of hay. It is easily planted in drills, 3 feet apart, one kernel every 6 inches, by an Eclipse Corn Planter, which plants 500 pounds of fertilizer in the drill, at the same time covering it and the corn, and rolling it all, at one operation, at the rate of 4 acres daily.

The best results I have had in crops have been obtained by using J. A. Tucker & Co.'s Bay State Superphosphates, 500 pounds being spread broadcast upon small loads of manure by a manure spreader, harrowed in, 500 pounds also in the drill, as above stated; and I will here add that one of the best things about the Eclipse Planter is that every kernel of the corn comes up evenly, and the crows will never pull up any of it.

About the seed I plant, I have had the best results from C. H. Thompson & Co.'s "Mammoth Ensilage" and the "Red Cob Ensilage" from St. Louis. When there is a good opportunity to market sweet corn, the best of it can be selected for market, leaving the forage in good condition for the silo. The best variety I have found for this purpose is "Stowel's Evergreen."

In estimating the capacity of the silo, after the forage is heavily pressed, a cubic foot will weigh about 50 pounds, usually a trifle under, so that it is very easy to calculate how much to plant, how much to feed, and how long it will last.

The "Dr. Bailey's Ensilage Cutter" will cut and elevate from 40 to 100 tons daily, with a 6 to 8 horse-power engine and boiler. The corn can be harvested and put into the silo in almost any weather, hot or cold, dry or wet (unless it rains too hard to work in the field), with less trouble, in less time, more security, and greater surety of perfect preservation, than any fodder crop can be harvested in any other way.

We have the statement of M. Goffart, who has tried it for many years, that Indian corn can be raised continuously, year after year, upon the same ground, by spreading upon the manure piles, each week, 100 pounds of ground bone to the equivalent of manure used upon an acre of land. I give my authority for this statement, because I have not tried it in this way. M. Goffart also states that he raises about 40 tons of fodder corn to the acre, upon land fertilized in this way, upon the same land continuously, and the forage keeps his cattle in perfect health year after year.

FERMENTATION IN SILOS.

Chemistry teaches us that fermentation takes place in the following order: first, the saccharine; second, the alcoholic; third, the acetic; fourth, the lactic; then a variety of other fermentations, either in quick succession or found to exist at the same time in the same substance. These transformations are accompanied with heat. At the fourth change, the heat is generally above 86°, and germs of bacteria are developed, and we have true fermentation, with continued evolution of Carbonic and Acetic Acids, in connection with a variety of putrid fermentations. These continue with rapid decomposition and recomposition, with increasing heat, until the mass goes to destruction, more or less quickly.

In silos, these germs of bacteria are supposed to get into the silos with the air, at the time of filling. They develop very rapidly, and multiply indefinitely, by subdivision. The germs will germinate into living activity at 86° of heat, and will germinate after exposure to a heat of 212° for some hours; but the developed bacteria will be killed at a temperature as low as 122°. Bacteria live upon oxygen, which they may get from the air, or they may get it from the sugar and starch in the corn, direct, without air. They live and thrive in an atmosphere of carbonic acid.

Now, with this explanation, how is it possible for corn to be placed with safety in a silo slowly, when mixed with all the air possible to get in with it, heating in the centre enough to kill bacteria, and toward the sides the proper temperature to develop germs into the greatest activity, the bacteria in the mean time multiplying indefinitely by subdivision in the best medium, sugar and starch, for supporting their life,— I ask how is it possible to stop such fermentation before the contents of the silo spoil?

On the other hand, suppose the heat does not rise above 86°, true fermentation does not take place, but the action of the air upon the forage, with moisture, develops a fungus growth upon the outside of the forage, which may continue, passing through mould and black rot to destruction. This often happens in corn fodder when the process of desiccation has been imperfectly performed, but true fermentation in the silo evolves and often ends in a light or dirty yellow residuum, with foul odors, more or less pronounced, nauseating and offensive. These conditions are usually found after heat and fermentation, just in proportion to the amount of air taken into and retained in the silo.

I have endeavored to give a *rationale*, as I understand it, of the process of fermentation found in the silo. But in my practical obsertions I have found that, as quickly as I could fill my silo, *Carbonic Acid* was also there, in *quantity*, the morning after the first day, and *Acetic Acid*, in *quantity*, the morning after the second day. My natural senses did not detect the presence of the saccharine or alcoholic fermentation. I did not get up in the night to call the roll, but found the substitutes in the morning, and have never since seen the delinquents to know them. I don't propose to contradict science, but I do propose to apply and use it according to my experience, and the HARD, COLD FACTS which have confronted me. I *know* that, if I get all the air out of my silo, I do not have heat or fermentation, consequently no loss of fodder and no foul odor; and I have come to look upon *Carbonic* and *Acetic Acids* as MY FRIENDS, in consequence of their early calls and assistance, in helping me to develop and perfect my system of perfectly *Preserving Green Forage* in its best condition.

THE SILO GOVERNOR.

Whenever forage is pressed in a silo, it packs where it is most dense, and becomes so hard that the air can neither get out of the corn nor out of the silo. It therefore remains in, and is pressed into the forage, which causes it to heat and ferment; it also prevents the corn from settling as it should, and acts as an air cushion, which causes lateral pressure upon the silo walls, and prevents settling enough to get juice at the bottom, and bringing it throughout the mass to the top. We therefore lose the great benefit of having a quantity of free juice in the silo, which benefit consists in reducing the temperature, making the forage soft and pulpy, rendering it more assimilable, and greatly increasing its feeding value. After Carbonic Acid has performed its office of displacing the air from the silo, it is absorbed by the juice, causing a partial vacuum, which causes the juice to rise gradually to the top, and is kept there, under pressure, by absorption and capillary attraction.

These operations are all brought about and controlled by the silo governor. Its action commences on the first day of filling, and goes on continuously for about two months, when the silo is ready to be opened. Briefly stated, it collects the air from all parts of the silo, conveying it to the outside. When the Carbonic Acid appears, being heavier than air, it sinks to the bottom as it permeates the forage, displacing the air, which it does gradually and quietly, without mixing with it: the silo governor also conveys the surplus quantity of Carbonic Acid outside, in the same manner; it also operates in the same way with Acetic Acid. These two acids and air are the only gases we have to contend with when we use the governor, which so perfectly removes and governs them that we never have heat or fermentation; consequently, no decomposition or development of foul odors. We keep a thermometer in the centre of the silo, and examine it frequently: we also measure the quantity of juice in the bottom of the silo daily, or as often as is necessary, by running a long stick down to the bottom of the perpendicular pipe of the lower governor. We get all the juice wanted from the corn, allowing it to accumulate on the bottom 20 to 30 inches deep. This last season, we had 6 inches of juice before we could put on the cover to press; the year before, we had 2 inches. This year we had a surplus of juice, and have been feeding from 50 to 100 pounds daily, mixed with the shorts, to our milkers. This juice was drawn off, clear, sweet, and odorless, from the bottom of the silo.

The governor collects and distributes the juice to and from all parts of the silo, and conveys the surplus from the centre to the outside, under the cemented bottom, to be drawn off as wanted. In making ensilage where no governor is used, it is seldom that any juice collects in the silo, even with 1 to 200 pounds' weight upon each square foot of surface; so that pressing green forage by this system requires not half the pressure to produce better results, with much greater economy. In controlling the operations inside the silo, we are guided quite as much by the quantity and quality of the juice as by the gases, odor, and temperature.

Carbonic Acid is perfectly wholesome in the stomach, and performs a good use in the silo; but it is necessary to be very careful working in a silo where it is, as no breathing animal can live in an atmosphere of it more than a few minutes. Acetic Acid is also very plentiful in the silo, and quite wholesome. It is the acid that causes the sour taste in every silo; and we are apt to get too much of it, as it is readily absorbed by the juice. But a great deal of it is taken out, through the governor, in a gaseous or vapor state. When this acid remains in a silo that has had no heat or fermentation in it, it is quite pure, and renders the food more palatable; but, when fermentation is present, it becomes decomposed, loses its acidity, and assists in producing foul odors, with a nauseous, putrid smell and taste. This state of things, more or less pronounced, is what constitutes the difference in the quality of ensilage; and its effects are noticed in the taste and smell, the foul odor imparted to the silo and barn, and upon the hands and clothing. Even the small quantity which the cattle can eat of it produces a nauseating effect; and the bad effect produced by it in milk, cream, and butter, especially when fed to delicate children, is positively unhealthy, not only to them, but to man or beast. Such a condition may be easily and entirely avoided in the preservation of green forage, and never exists where there is no heat or fermentation, or where the governors are used, in a good silo, to prevent it.

With good smooth walls, held up with iron rods built into them, with the governor to take out the air and gases, we have but little lateral pressure; and yet we bring immense vertical pressure to bear directly and uniformly, which condenses the forage, without impacted strata in the mass, giving us results, without fear of accident, not obtainable by any other means.

CUT OF SILO GOVERNOR.

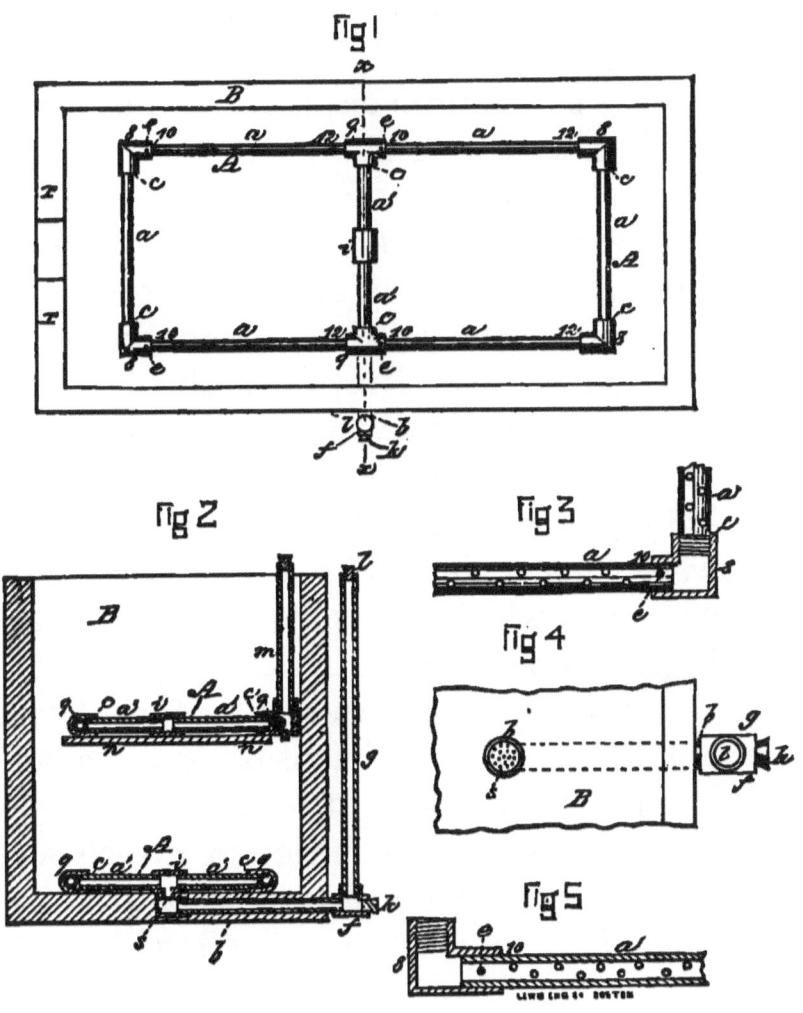

DESCRIPTION OF THE SILO GOVERNOR.

Figure 1 is a top view of a silo ready to receive the ensilage, and showing a portion of my apparatus resting on the floor.

Figure 2 is a vertical section cut on the line xx of Figure 1.

Figure 3 is a portion of pipe, on an enlarged scale, taken from one of the front corners of the apparatus, and placed bottom side up to show the air holes on its under side.

Figure 4 is a top view, on an enlarged scale, of a portion of the bottom of the silo before the principal portion of my apparatus has been placed in position, showing the upturned end of the drip pipe, and the strainer in its mouth.

Figure 5 is a section of portion of pipe and elbow, on an enlarged scale, showing the wooden peg which prevents the pipe from turning in its bearings.

I construct my apparatus as follows: I take iron pipes a a' of any dimensions desired, and join them together so as to form a frame A, with a continuous air connection (which also communicates with the drip pipe b), by screwing each of the ends of the pipe a and a, into its connecting elbow 8 or coupling 9, as shown at *c*. The ends 10 and 12 of the pipes *a* are thrust or telescoped into their connecting elbow 8 and their couplings 9, as shown in Figures 1, 3, and 5. All the horizontal pipes, except the drip pipe *b*, which runs toward the side wall of the silo B, are perforated on their under side with holes about one-fourth of an inch in diameter, and about six inches distant from each other, as seen in Figures 3 and 5; the ends 10 of the pipes *a* are each held in position by a small wooden peg or pin *e*, as shown in Figures 1, 3, and 5.

A T-coupling *f* is screwed on to the projecting end of the drip pipe *b*, and the vertical pipe *g* is screwed into the upright branch of said coupling *f;* the vertical pipe *g* affording an outlet or means of escape from the silo for the air and gases. The pipes *a a′*, which extend transversely across the centre of frame A, from side to side, are screwed into the couplings 9, 9 and *i*, the downward branch of the latter coupling fitting loosely within the upturned end of the drip pipe *b*. The water, juices, etc., from the forage, are drawn off when desired through the drip pipe *b*, the outer end of which is provided with a stopper *k;* but a faucet may be employed instead of the stopper, if preferred.

The mouth of the vertical pipe *g* I close with a stopper *l*, or with a cap.

If the silo has a capacity of over 150 tons, or is more than fifteen feet deep, the apparatus for the bottom of the silo being in place, the cut corn is piled upon it in the usual way; and, when the silo is about half full, another apparatus, or frame A, not differing materially from that on the bottom of the silo, excepting that its vertical pipe *m*, which performs the same office as the vertical pipe *g*, and also has its mouth closed by a stopper, rises from the inside of the silo, as seen in Figure 2.

For the better support of the second frame A, and to prevent the small holes at the bottom of the pipe from becoming stopped with the forage, it is placed on a skeleton platform *n*, Figure 2, composed of narrow strips of furring; and upon this platform *n*, with the frame upon it, the cut corn is piled until the silo is filled. To secure the second frame A in its place on the skeleton platform *n*, nails or staples are driven into the platform for that purpose. As soon as the silo is filled, the ordinary planks are put upon the top of the forage, and the weight placed on said plank. Nothing should be put between the forage and the plank.

When the freshly cut corn is placed in the silo, it has not yet had time to become much wilted, if, indeed, it is wilted at all. Consequently, the air which remains in contact with it there, is in a much freer condition than it is after it has wilted; for through the operation of wilting the said air becomes much more intimately associated with it, and much more difficult to separate from it. Therefore, during the process of filling the silo containing my apparatus, a large portion of the air in contact with the forage will be taken into the pipes *a a'*, and escape into the surrounding atmosphere through the vertical pipes *g* and *m*.

The parts of the governor are now made to be screwed together, which is found to be preferable to "sleeving" them together, as was formerly done. Right-hand screws are used at every joint, excepting the right and left couplings at the cross-sections.

* We take the perpendicular pipes from any part of the governors. When used to take temperature, they are taken up through the centre of the silo. The cut in Figure 2 represents a pipe taken up at one side, about 6 inches from the wall.

In opening the silo to remove the forage,

when the first part of the frame A of the apparatus is reached, the pipe a nearest to the front end r of the silo B is pulled away from its connections with the adjacent longitudinal pipes a, and the latter are then also removed, the pegs e, which held the pipes a in position, having been first taken out, or broken off.

As the work of discharging the silo proceeds, on arriving at the central portion, the pipes a', with their couplings 9, 9, i, after pegs e have been removed, can be pulled out and lifted from their place, and their adjacent longitudinal pipes a drawn out, leaving only the pipe a at the rear part of the silo to be removed when reached.

In building a new silo, I place the drip pipe b, Figure 2, so that its upturned end will be flush with the surface of the bottom of the silo. Just below the surface of its upturned end, I place a strainer s, Figures 2 and 4, which will allow water and juice to pass freely, but will arrest coarse pieces of matter.

When liquid rises in the pipe g, it can be drawn off by removing the stopper k.

The silo governor has air passages within three to four feet of every part of the forage to the outside of the silo, from which to discharge the air, not only after it is packed, but while it is being filled.

Air can be taken out of the silo in larger quantities from the bottom and central parts of the silo than can escape from the top.

While the air is going out of the silo, there can be no ingress; and, as soon as egress ceases, the air passages should be closed, by stopping the mouths of the perpendicular pipes *g* and *m*.

The governor will take off the lateral pressure from the walls. There will be nothing like the pressure of an air cushion in the nature of hydrostatic pressure. But there must be weight enough upon the ensilage to press it to such a degree that it may be cut vertically from top to bottom, leaving a smooth, solid front to prevent the ingress and action of the air.

A HALF EMPTY SILO.

The following cut is introduced for the sole purpose of showing the situation of the two governors when one-half the ensilage has been removed. It also shows the vertical pipes.

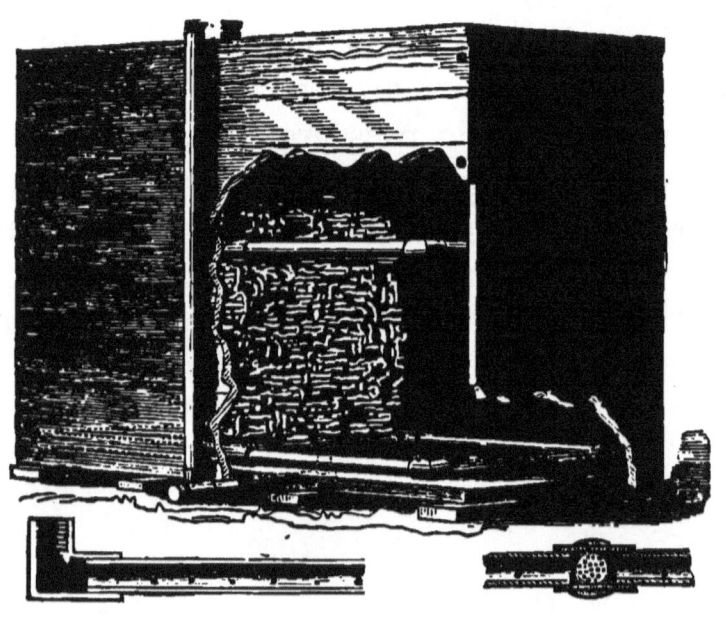

SILAGE *versus* DRY FODDER.

Professor Arnold, being asked why three tons of good silage have a feeding value of one ton of the best hay, replied that in green succulent foods the cellular tissues have not been converted into woody fibre, and in mastication and digestion all of the nutritive substances in these cells are quickly acted upon by the saliva of the mouth, and then the gastric juices of the stomach, and all the nutriment is assimilated with only a minimum expenditure of force by the animal economy to digest it. The natural moisture of the plants, when green, also acts as a compensation, and requires but little beyond the gastric juice to make the food fluid enough for digestion. With dry food, nature is heavily taxed at all points to make good the loss of the juices or moisture of the food. The secretions of the mouth are called upon to moisten the dry food. The woody fibre of the plants must be broken down and disintegrated by the power of gastric juice to set free the real nutriment of the food. This force is several times greater than is necessary when succulent food is fed. All this extra expenditure of force must be supplied by the animal, and therefore calls for an increased amount of food to make good this demand, or else the animal falls off in flesh.

EXPERIMENTS WITH ENSILAGE.

At the annual meeting of the New York State Agricultural Society, recently held at Albany, Dr. E. L. Sturtevant, director of the State Experimental Station at Geneva, read a paper on ensilage, an abstract of which we give herewith:—

In the experiments carried out last year at the station, sweet food, purposely acidified with a measured quantity of acetic acid in about the same proportion as analysis showed to exist in ensilage, gave better results in milk and by live weight than did the same food without the acid; and the doubling of the acid ration was followed by an increased improvement in quantity of product. A careful examination into the kinds of food fed during the various periods showed that one apparent effect of the acid was to improve the appetite of the cows and cause them to eat a slightly larger ration than they had been using previously. We are thus led to believe that, so long as the acid fed is not in a proportion beyond proper condimental relations, it is a valuable adjunct to food. When we notice that the use of ensilage as sole food has not produced a satisfactory condition in the animals thus fed, we fully believe that the feeding of the acid beyond its condimental proportions is not advisable. We are led to believe that ensilage must be considered as a valuable food when judiciously fed; and in the hands of a judicious feeder it may possess a value superior to that of the raw material, inasmuch as it contains the same amount of nutrition in addition to a certain condimental effect upon the animal.

THE OPINIONS OF EMINENT AGRICULTURISTS.

The following certificate from gentlemen eminently and reputably known for their practical knowledge of agriculture and the application of business intelligence in their operations will be interesting in connection with the subject of ensilage. It reads as follows:—

"The undersigned, having made and fed ensilage for several years, believing that we have arrived at certain and uniform success, offer to those who are in doubt, this

CERTIFICATE.

"This certifies that we are making ensilage *without heat or fermentation*, in air-tight silos, cutting the corn in one-half to three-quarter inch lengths, weighting 100 pounds to the square foot, and with this pressure getting one foot or more of juice in the bottom of the silo. We remove the air from the silo by using Colcord's Silo Governor, which causes a heavy vertical, with very little lateral, pressure. We obtain as uniform results cold, moist, soft, and pulpy ensilage, of the natural color of the corn, without offensive odor, imparting no odor to the silo, barn, hands, or clothing, but often of a bright, sweet smell, and sometimes the odor of nice, dry corn fodder. We feed an average of 60 pounds daily to each cow, and our cattle eat it all without any waste.

"We regard Mr. Colcord's system as the true, if not the only true, method of ensilaging green forage crops, and recommend it as sure, uniform, economical, and less troublesome than any other. By using this system, with the governor, according to directions, any one may be sure of success with ensilage.

"EDMUND M. WOOD, Boston, Mass.
"T. E. RUGGLES, Milton, Mass.
"CHARLES L. COPELAND, Milton, Mass.
"C. A. DAVIS, Natick, Mass.
"BERNARD MONAGHAN, Dedham, Mass."

[From the *Farm, Field, and Stockman.*]

SWEET FORAGE IN WINTER.

COLCORD'S GOVERNOR IN PRACTICAL USE.—PARTICULARS FROM MR. COLCORD HIMSELF.

General C. H. Howard:

Sir,— I never planted corn any better, cultivated better, or manured better than this year; but the season here has been unusually bad for corn,—cold and wet to a degree I have never seen before. The corn did not grow. It was two feet less in height, the leaves were about half the usual size, and the stalks small: it was badly frost-bitten about September 1.

During the months of September and October, the rain was as continuous as the rainy season of California or some parts of the South. Very little corn around here ripened, and nearly all the fodder spoiled in curing.

I cut mine into the silo September 18 to 24. It rained all that week except half a day. We were four days putting it in. It rained so hard 2 days that we could not work. It averaged 13 tons to the acre (last year 19 tons). All the corn was in the milk. It was put in very wet, cut to half-inch. I put in 3 governors, 1 on the bottom, 1 in the centre, and 1 inverted immediately under the plank covering, because I used the same splined cover that I did last year. When the silo was ⅔ full, there were 6 inches of juice all over the bottom: no carbonic or acetic acid this year. Last year both were very abundant. The juice is sweet, but would test slightly acid. The temperature in the centre of the silo is 76° to 78° (last year 72°). There are 18 inches of juice in the bottom, (last year 30 inches). I can get any quantity I want at any time.

The top is kept perfectly level with ten jack-screws. It is under perfect control, and no trouble to press it. I can discover no foul odors, and think it is ripening or curing very nicely.

I hope to be able about December 1 to send you a sample of the juice and a sample of the preserved forage: if I do, I shall press out the juice from 2 or 3 pounds of the forage near to the top of the silo, bottle it, and pack it in the identical forage I press it from. I don't understand why I should have no acid in it this year and so much last year, and what the difference will be when I come to open the silo. If this turns out good, it will settle the question of being able to be sure of the corn crop every year. It seems impossible to have another year as bad as this; and I could see no objection to the acid I had in it last year, and all my experiments went to prove it. I shall weigh all my stock when I begin to feed it, and every thirty days after, and note the difference in milk and flesh. The difference must be in the quality of the crop, although there is 6° higher temperature this year. I am trying to find out about that capillary attraction. Last year the juice went up about 1 inch a day after I ceased to press it: this year I am not pressing so heavy, and it doesn't appear to rise so fast. My last month of feeding last year showed quite an increase in feeding value.[*]

MY EXPERIMENT SILO.

In building my silo, I took nearly level ground, laying it out to build an air-tight pit 12 x 32 x 20 feet, excavated 5 feet, putting in the foundation of cobble-stone 20 inches wide, 18 inches below the bottom of the silo, with a 4-inch land drain around the outside, pouring over the foundation thin, mixed cement. After it became firm and level, I erected a

---

[*] This article was published soon after the silo was filled. Later on, carbonic and acetic acids put in an appearance.

staging 18 feet high. The upright timbers next to and on both sides of the wall to be built were 6 x 6 spruce timber, 5 feet apart, securely fastened across the pit by 2 x 8 plank, sawed exactly to one length, securely fastened near the top, bottom, and centre by one 6-inch round spike in the centre of each 6-x-6-inch cross connection, to give a perfectly flat, firm bearing. The planks running lengthwise were fastened to the uprights in the same manner, care being taken to have the uprights perfectly perpendicular, without variation in distance between the walls. To insure perfect accuracy, the inside uprights were spiked together and raised in pairs. Straight timber was selected, and as far as possible straight-grained; the diagonal braces of 1 x 6 fence boards were used.

After the inside staging was up, the outside timbers were raised and placed opposite the inside ones, leaving space for the 18-inch wall and the plank on each side, also for laths $\frac{1}{4}$ inch thick to be placed between the plank and the uprights, to be taken out each time the plank is raised. The inside and outside upright timbers, also the outside 4 x 4 staging, were securely fastened together with 1 x 6 fence board, 6 feet long. Thus the inside and outside stagings were securely fastened together, the connections being sawed away as the wall was built up between the stagings. The wall planks were 18 inches wide, planed to even thickness. These were placed all around the pit and mitred at the corners. Between these planks the wall was built up daily from 12 to 16 inches.

When the inside staging was removed, there was not a variation of $\frac{1}{8}$ inch in the length of the walls from top to bottom or from end to end.

I was thus particular in building, because I was trying to make an air-tight pit, in which I could exhaust, which was equal to packing 384 square feet of covering tight enough to exhaust 6,528 cubic feet, which every one said I could not do.

The mortar was composed of one part cement, two parts coarse sand, two parts small, clean cobble-stone, two parts small broken stone, and water in the proportion of about 30 gallons to each barrel of cement. This was taken to the pit in buckets, poured in and packed with trowels, to keep the stones from the plank. Iron rods $1\frac{1}{4}$ inches in diameter, with strong flanges at the bottom, terminating 4 feet above the top of the wall, with long screws and double 1¼-inch nuts, were built into the centre of the side walls from the foundation up. These rods should be placed about 8 feet apart, commencing 4 feet from the end walls.

Eight by eight spruce timber was used to connect opposite rods across the top of the pit; 1½-inch holes were bored through the timbers, to allow them to slide freely on the rods; cast-iron washers 1¼ x 6 inches were used under the nuts. When the wall was up half-way, all around level, 2 x 6 planks were set up on the outside of the wall between the 6 x 6 upright timbers and the building plank, for the purpose of setting the upper half of the wall in 6 inches all around the outside, leaving the part yet to be built 12 inches thick. In the 6-inch ledge at the bottom, reserved to build upon, was placed a strip of 2 x 2 all around to fasten the woodwork to.

Before we commenced laying up the wall, a drip or drain pipe 1½ inches in diameter was placed at the bottom, from the exact centre of the pit to one of the corners, to come out into a well 3 feet deep. This outside end was placed 8 inches below the bottom of the pit, with a ⊥ turning up 3 inches inside the face of the wall. Into this ⊥ was screwed an upright 1¼-inch pipe, which was built into the wall, coming out at the top of the 6-inch ledge upon the barn floor, terminating 2 feet above the floor with a plug and side stop-cock, arranged to collect the gases for examination, to sound the depth of juice and draw it off at the bottom over the well, also for general purposes of examination. The end of the drip pipe at the centre of the pit terminated in an elbow, with 3 inches of pipe coming up through the cemented bottom and flush with it. These pipes are a part of the governor, which in this pit is a frame of 1-inch iron pipe 26 feet long and 6 feet wide, perforated with ¼-inch holes every 6 inches, arranged to sleeve together at the corners, in the centre, and at the sides 6½ feet from the ends, and put together in such a way that the whole of it can be put in in fifteen minutes, and every part of it is taken out separately, as it is found in taking the forage from the pit; it is also placed so that the forage cannot stop up the ¼-inch air holes. In the centre of this frame is a ⊥ that turns down 3 inches into the drip pipe. In practical use, this is all of the bottom governor.

The governor being in place, we cut the corn in ½-inch lengths, fill the pit a little more than half full, level and tread it evenly; then upon a skeleton wooden frame we put a second governor, in all respects like the other, except, instead of turning down the outlet into the drip pipe, we turn it up, with a perpendicular pipe coming up through the forage from the centre of the pit and through the plank covering, and terminating, like the other, with a plug or cap. When the pit is full and trod down evenly, cover with 2-inch spruce planks fitted to slide down the walls nicely and evenly; press it firmly enough to get about 2 feet of juice in the bottom, and in practical operation this is all that is required. But in this pit, which was made to try any experiment and test any principle in the direction of possibilities, or perfect preservation, a third governor inverted was put in, and the three governors were sleeved together, having a continuous outlet from top to bottom, closed with stop-cocks and plugs, having 432 outlets for gas and air from the forage into the pipes, distributed evenly through the mass. The top governor was laid directly upon the corn, the pit was covered with 2-inch splined planks, accurately fitted so as to slide down the walls as the mass settled. This cover was covered with two layers of thick paper, and a 4½-inch rubber packing all around the walls, making it air-tight. This was also covered with 4 inches of damp sand.

Two lines of 6 x 8 timber were placed the length of the pit upon the cover, upon which were placed 2-inch jack-screws under the 8 x 8 timbers through which the iron rods passed. This arrangement, by reason of the elasticity of the corn and the springing of the timbers above the jack-screws, gave a continuous pressure, which was found to be ample and safe. In this way, I was able to get an air-tight exhaust. All the air and gas had to come out through the governor, giving an opportunity for examination and taking the temperature daily at different depths. At no time was the temperature above 72° in the pit, which was about the outside temperature when we commenced to fill. Carbonic acid appeared in the pit the morning after the first day we cut, and the next day acetic acid put in an appearance. These, with air, were the only gases or vapors that have come out of the pit; and these appear perfectly pure, without any odor,—something I have never seen before in any silo. In fact, there has never been any heat, fermentation, or foul odor in the pit. Juice drawn from the bottom is odorless; and when, by long exposure to the air, it does change, it turns to bright, odorless vinegar. There is no odor of ensilage in the pit or stable, and not any waste in the pit or at the feeding-troughs, but is all eaten up clean. I am now feeding an average of seventy pounds daily to each animal. Many of them would eat considerably more. I think, however, that I shall find a variety for cows is better than any one kind of food.

The bottom and centre governors took the air out so fast while we were filling that before it was full we had 2 or 3 inches of juice all over the bottom of the pit. There was very little lateral pressure until after we began to press. The 1-foot walls were found to be strong enough, the strain upon the iron rods preventing any fear of their pressing out. One of my aims was to get the juice as near the top as I could, to make the mass soft and uniform throughout. I have succeeded perfectly in doing this, in getting an exhaust.

The mass is now cut down vertically $13\frac{1}{2}$ feet, and back across the end 10 feet, with a hard, smooth face which does not change, let the air in, or the juice down. I can take a handful of the forage and squeeze the juice from it, from any part of the face. There is no air in it, it remains sopping wet and cold from

Preserving Green Forage

the top plank to the bottom. To test the exhaust, ·I connected the governors both at top and bottom of the pit with a steam vacuum pump. As soon as the air was removed and juice came into the pump, I cut off the connection with the bottom, and the pump threw a stream of juice from the top 10 feet high into the air. Therefore, more pumping or pressure was useless, the juice has taken the place of the air, and capillary attraction is keeping it there throughout the mass. In practical operation, capillary attraction is sufficient to convey and keep the juice at the top. This has the effect of keeping the forage wet and cold, and seems to give it a ripening process, rendering it more palatable and assimilable, as evinced by the continual improved quality as we cut into it. The cover should be removed no faster than we cut down vertically. As I have had no heat or fermentation in the pit, the forage does not heat up when taken out and exposed to the air. In very cold weather, I pour over the forage in the feed-box one to two gallons of hot water to each 100 pounds; but it will not start to increase the heat, as there are no germs of fermentation in the forage, apparently. The forage has a density in the pit of 50 pounds to a cubic foot.

The great advantage in this manner of pressing with screws is that we get the amount wanted wherever and whenever we want it, it can be put on and taken off at pleasure, the elasticity of the corn with the spring of the timbers above the jack-screws gives it a continuous pressure, the time, trouble, and expense of weighting are entirely avoided. Another advantage is, the iron rods can be carried up 6 or 8 feet above the top of the wall, arranging the studding above to receive the planks and virtually building the silo so much higher. When the pressure is put on, it brings the cover down between the cement walls; and these planks, also the covering planks, can be packed away above the pit, just where they will be wanted for future use.

BUILDING SILOS.

The door in the end of my silo is 4½ x 6 feet. The door-frame is made of 3-x-12-inch plank. All around the outside of the door-frame, I spiked a strip of 2-x-4-inch plank. This strip, being embraced in the masonry, formed a lug, or shoulder, to hold the door-frame in position, and prevent its being disturbed. I anticipated some trouble in making the doorway tight; but, with a 2-inch matched plank, made to fill the opening in the door-frame, and by fastening with 4-inch lag-screws a 2-x-10-inch plank on to the 3-x-12-inch door-jambs, I succeeded in rendering the doorway absolutely air-tight. And it remained so until the silo was opened.

I used a very heavy old cart-wheel iron tire to bind the end of my silo together, over the door, by laying the tire upon the top of the door-frame, turning the ends up 6 inches into the side walls, and passing a $\frac{3}{8}$-inch iron bolt through the door-frame, the iron tire, the walls, and the cap of the silo, turning the nut firmly on the bolt.

I feel that I have fulfilled my promise in making these experiments; and by them, with the experience gained in five years past, I am able to prove every statement I have made about this system, and now have the results to show. I think I may congratulate the stockmen in this country upon the certainty that green forage can be perfectly and uniformly preserved year by year, at a cost of about one-third the feeding value of hay; that it costs no more to handle than hay in time or money; that it is a surer feeding crop to raise than any other; that the insurance risk is far less from lightning, fire, flood, or drought, than upon any other stock food; that cattle drink only about one-half as much as when fed on hay; that by this system we can feel insured against cold or hot, wet or dry seasons, and silos can easily be protected from frost. My silo is boarded up outside with 1-x-6-inch rough feather-edge boards to the top, and packed between with 6 inches of sawdust on the exposed side and end. It will be found to be the most economical way to build good, tight silos. The expense of building them depends somewhat upon the soil and situation: the expense should be about the same in building for stock fed with preserved forage as for hay. There is no economy in building cheap silos, and poor economy, with great waste, in packing corn in whole, filling slow, and having heat and fermentation that cannot be controlled.

No lime should be used in the mortar for building silos. Any one intending to build a silo, or having one that is not satisfactory, will do well to call and see this one. It will be a great satisfaction, and a saving of time and money, as the important experiments have nearly all been made, and are open to any one that calls. It is very easy to see how silos can be altered to make this kind of forage.

In building a silo, it pays to build well; and, to prove all the facts about it, I tried to build perfectly, in order to test the possibilities of a perfect system, and have inserted in this book a picture of the staging, inside and outside the silo wall, when it was 5 feet high, taken from a photograph (see cut, p. 68). When building my silo, the weather was 100° in the shade, the plank between which the wall was made was exposed to the sun on one side, and cold, wet, grout cement mortar on the other side. The consequence was that the plank swelled, or rolled, between the upright timbers ¼ of 1 inch on each side, making ½ inch variation, in some places, between the side walls; and this was and is the only deviation in the walls.

The following cut, taken from a photograph, shows my silo in process of building:—

SILO IN PROCESS OF CONSTRUCTION.

DIRECTIONS FOR PUTTING IN AND REMOVING THE SILO GOVERNOR.

The governor is made of 1-inch pipe, with holes ¼ inch in diameter and 6 inches apart, along one side of the pipe. The pipes are arranged in frames around the inside of the silo, about 3 feet from the walls, with 1¼-inch pipe across the centre of the frame. In the centre of this pipe is a ⊥ to turn up, in which to screw the perpendicular pipe, if it is to come up in the centre of the silo through the forage. In the centre of the bottom governor, the T turns down to enter the drip pipe, which runs from the centre of the silo, declining about 6 inches, to any convenient place outside the wall, under the cemented bottom, the outside end of the drip pipe terminating with a stop-cock, to draw off the juice; also, a ⊥ in which to screw the perpendicular pipe which comes up outside the wall to any convenient height, usually to 2 feet above the barn floor, for convenience in measuring the juice and to have the stop-cock and outlet under cover, to avoid frost.

These bottom governors can be arranged to have the outlets on a level with the bottom of the silo, by cementing a pipe in the wall large enough to sleeve the governor into. The stop-cock, from the drip pipe in my silo, comes out over a little well about three feet deep, to permit a pail to be held to catch the juice. All the governors, except those lying directly upon the bottom of the silo, have the perpendicular pipes come up through the forage. All the horizontal pipes should be placed upon strips of board about 4 inches wide. All the governors should be carefully put in, so that the ¼-inch holes will be at the bottom, which makes a passage the whole length of the frame for air and gas to get into the pipes. In putting in the centre governor, the forage should be well trodden level, or it will get out of place, and the pipes bent in settling.

Sometimes, in old silos, when people don't want to go to the trouble and expense of making alterations, I put in both of the perpendicular pipes from the governors, to come up through the forage to the top, in which case, if there is much juice in the bottom, it must be pumped out before cutting down the forage into it, because air coming in contact with the juice is apt to change and injure it, and carry the injury into the forage.

The governors, as we make them now, are much cheaper than formerly, and work equally well. All the connections are made with right-hand screws, excepting the right-and-left-barred, or ribbed, couplings on the sections that go across the frame, so that, in putting the governors together, lay the side pipes down as they go, and screw them together. Take your centre 1¼′ in cross-section, unscrew the right and left coupling, and screw each half into the opposite sides. Do the same at both ends. Then turn these two halves bottom-side up, and see that all the ¼-inch holes are on top. Then turn them over again, screw the right and left couplings together, then your perpendicular pipes. The frame is then ready to be placed in the centre of the silo. See that your centre governor is placed a little more than half-way up, fastened to the skeleton platform with staples, or plumbers' fastenings. When you again find it, in removing the forage, it will be more than half-way down. Remove the governor as you come to it, unscrewing the right and left couplings first, then each part in the manner you put the frame together, sawing or breaking away the skeleton platform as you come to it. All the connections are made, and disconnected, by using ordinary gas-pipe tongs. They are very easily handled, as all the screws are right hand, except the right and left couplings, and do not require to be screwed very tight.

When the governor is in place, there is a row of ¼-inch holes on each side where the pipe comes in contact with the bottom of the silo, or the strips of boards, so that, when the fodder is placed upon them in the silo, it falls down on each side of the inch pipe, leaving a pathway for the air and gases to pass along to the holes in every part of the frame.

Supposing that your silo inside is 12 x 32 x 20 feet, your two governor frames would be 6 x 26 feet each. There will be a distance of 3 feet all around from the wall to the governors; also, 3 feet from the centre of the silo to the governors. You place your upper governor 11 feet from the bottom, and fill the silo 20 feet high. Covering and pressing may reduce the mass to 12 feet. You then have 3 feet from the top to half-way to the upper governor, then 3 feet to the upper governor, then 3 feet to half-way between the two governors, then 3 feet to the bottom governor, bringing every part of the forage to within about 3 feet from these rows of ¼-inch holes in the pipes and the openings between the top plank, giving ample opportunity for the escape of gas. In this way, while filling a silo of this size, in four days, I had 6 inches of juice in the bottom before I could get the cover on or give it any pressure; while other silos, having no governors, had no juice upon their bottoms, even when weighted with from 100 to 200 pounds to each square foot of top surface. By this system, it does not require one-half the weight or pressure to secure the ordinary results gained in the management of silos; and I am led to believe there is no other way known as yet to prevent heat or fermentation, or one likely to be devised as perfect and economical as this.

74 Colcord's System of

TABLE-TOP CORN-CART.

DESCRIPTION OF THE TABLE-TOP CORN-CART.

The cut represents a level, portable table, designed to take the corn from the field to the cutter, to be cut without laying it upon the ground, avoiding handling, soiling, getting it gritty or any mixing it with stones, gravel, wood, or dirt of any kind, to dull or injure the knives, or to get to the animals in their food. The top is 6½ x 9½ feet. The wheels are 40 inches in diameter, with a 4-inch tread. It is arranged to use one horse or two. The tall corn is spread upon it when cut, the butts all on one side, and can be fed into the cutter directly from the cart, which is the same height as the cutter. The rollers of the cutter feed the corn to the knives by merely pushing the butt-ends to the rollers. In practical use, it works admirably, and is a very useful cart upon a farm for gathering fruit and other crops in the field.

FAULTY SILOS AND FAULTY MANIPULATIONS.

It is necessary to have tight silos, with smooth, perpendicular walls, the opposite walls to be equally distant from each other, in all places, in order to preserve green forage perfectly. It is also necessary that the walls should be strong enough to stand heavy pressure; also that they should remain tight while the forage remains enclosed in them, and not absorb the juice from the forage. These are the ends to be sought in building a good silo; and it is of comparatively small importance how they are made or of what material, if these ends are attained. I seldom visit a silo and find these conditions fulfilled.

If any one will examine his silo carefully, with a good, long, straight edge, and measure the distance carefully between the opposite walls with a rod just long enough to touch each end at the narrowest place, and have every covering plank cut ¼ inch shorter than his rod, he will realize the importance of the conditions above stated, and give attention to them, because the cover cannot be pressed down evenly if it binds anywhere upon the walls; and, wherever it binds, the air will get under the plank and spoil the forage. Another reason is that, if the walls are uneven, the forage as it descends under pressure will be crowded from the wall wherever a bulge exists, and, when it passes the bulge, it will let in the air. The same holds true in pressing the cover down evenly, which is one reason why pressing by screws upon timbers placed across the plank covering is so much better than weighting. If these things are carefully attended to, there will be no waste whatever on top or around the walls, unless the cover is allowed to come up when removing the plank for cutting down the forage. The pressure should not be removed from any plank excepting those that are removed for the purpose of cutting down. This removal is the cause of so much waste in so many silos. The governors will remove the air from the bulk of the forage, and prevent heat or fermentation, or any foul odor or damage; but air getting in from the outside will cause the forage to mould and produce black rot. The air does not get into the forage through the face of the perpendicular cut, if under proper pressure, and if cut down with a sharp hay-knife. The whole cover should not be removed from all the top at one time, as the action of the air gives it double the acidity to be found in the forage 2 feet below the top layer that has been forked over. These facts I know from personal experience, observation, and chemical tests. Whenever I have complaints about silos in which the governors are used, I try to ascertain the cause of the trouble. I have never found it to be in the governor, and have always been able to find the cause. In one instance, I found a large amount of liquid in the bottom; also, that the water-line in the earth all around the silo was precisely the same as inside. The silo had no drains around or from it. In another instance, I found the governor in a cheap wooden silo. In the centre of the silo the forage was perfect, but some quite large holes in the boards of which the silo walls were made had given access to air, which caused the forage to go to black rot, and left large vacant places in the forage, large enough for a man to lie down in. It was not weighted heavily enough. Still, it had no true fermentation in it.

Most of the troubles, I find, and they are quite common, are in light weighting, uneven walls, and covers that would not go down inside the walls, the planks being too long to pass the numerous large bulges. To one man I sent two governors to go into his two silos. He put both into one silo, none into the other. The defects in the walls caused in some places 3 inches of waste around the silo containing the governors. The other silo had about 1 foot of waste around the walls. His silo was too far away for me to visit, and I do not know whether he had true fermentation in either of them; but he was satisfied with the governors, and contemplates building more silos and ordering more governors.

There is another great trouble I have found, which is mixing my system, without heat or fermentation, with the opposite one of slow filling and light weighting, with heat. These two systems are incompatible, and should never be mixed or confounded, no matter who suggests or advises it. It has spoiled some of my best work, even after perfect forage had been made in all parts of the silo. It injures the forage to remove the pressure faster than the forage is removed.

I do not object to weighting if it can be done so as to press down the cover evenly, as could be done by placing planks across the cover and weighting upon them; but it would then be very difficult to remove the weight and plank from that part which is wanted to cut down without disturbing the other plank and getting air into the forage.

REMEDY FOR FAULTY SILOS.

Where silo walls are made of heavy stone masonry, they can be faced smooth by floating a coat of coarse sand and pure cement, frequently using a long, straight edge in every direction, until the depressions in the surfaces are all filled; or, in case the wall is very uneven, perpendicular timber could be firmly placed in front of the face, and a straight, wide plank placed between the timber and the wall, arranged to slide up evenly, and cement mortar (no lime) filled in between, daily, to the top.

If such walls are strong and well made, and T rails of railroad iron, 1 foot longer than the silo is wide, can be obtained, openings can be cut out of the bottom of the silo just large enough to admit the rail about 3 inches below the bottom of the silo, and under the walls about 1 foot on one side, and 6 or 8 inches on the other side, to allow the rail to drop in and be placed firmly, 6 inches under each wall. A piece of iron rod, 1 inch in diameter, is bent so as to form, midway between its ends, an eye large enough to receive a hook on the end of a long iron rod, 1¼ inch in diameter. The ends of said rod are bent so as to form a stirrup, or loop, large enough to slip over the railroad iron.

The said long iron rod, having a hook on its lower end, has a long screw cut on its upper end, to receive a double nut and 6-inch washer. The long rods are made to reach from the bottom up to from 4 to 5 feet above the top of the silo, and the upper ends of said long rods pass through 8 x 8 timbers, upon which the double nuts and washers bear for the purpose of producing the desired pressure on the forage. These long rods are placed along the side walls, directly opposite to each other, commencing not more than 4 feet from the ends of the silo, and not more than 8 feet apart. Of course, these long rods are better built into the centre of the walls, as they serve to help hold up the walls when under pressure. The long rods should be placed as close as possible to the side walls of the silo. I have put these long rods in, both ways, in the centre and outside the walls. I have ordered them of the Boston Bolt Company, describing them as bolts. They came just what I wanted, at very satisfactory prices.

The irons across the bottom should have a

good bearing under the walls. Otherwise, if the bearing is just under the edge, the pressure, together with the lateral pressure of the forage, may cause the wall to topple over.

[From the *Dairy World*, October, 1886.]

ENSILAGE AND ITS IMPORTANCE.

A NEW DEPARTURE IN PRESERVING GREEN FODDER.— THE TESTIMONY OF EXPERTS.— AIR PERFECTLY EXCLUDED.— THE FORAGE COMES OUT SWEET.

The nature of ensilage is so well attested and understood in Europe and America that no plea is longer necessary in its defence. The only question now to be considered is as to the best and cheapest means of preparing the fodder. It is not necessary to go over the means heretofore used to prevent fermentation. The intelligence of inventors has been directed constantly to the easiest and most perfect means of keeping fermentation within bounds, or of preventing its action unduly. The measure of success with silos is in the more or less perfect exclusion of the air. If this is perfectly accomplished at the time of filling the silo, there will be neither heat, fermentation, decomposition, nor foul odor.

The theory of filling the silo slowly and allowing the temperature to rise from 122° to 180°, to kill the bacteria, Mr. Colcord says, is a fallacy. The fermentation cannot be controlled. The ensilage is always sour first, and becomes sweet (that is, not so acid) by progressive fermentation, with foul odor, and always at the expense of the quantity as well as the quality of the forage. By the system here described, these changes do not occur. The forage is kept in its natural condition, as follows:—

"Sweet ensilage, as commonly understood, does not represent preserved green forage produced by this system. The term 'sweet,' as originally used, was not used in a sense as opposite to sour, but as opposed to putrid (as sweet meat).

"The average quantity of ensilage, as heretofore made, *that can be fed daily*, is about forty pounds. The cattle do not care for more; but forage made by this system and device can be fed sixty pounds or more daily, and all of it eaten without any waste, giving the best results, even better than fresh-cut fodder.

"The most interesting feature in this system

is its economy. From corn can be raised the heaviest and best crop of forage at the lowest cost. The big butts contain the most sugar and starch. By this system, these large stocks are preserved, and come out in a soft and pulpy state, and are all eaten. By those who have tested it by keeping accurate account, the average cost of preserved green forage is $2 per ton. In feeding value, three tons of it are equal to one ton of the best hay, making preserved green forage at $6 equal to hay that can be readily sold for $18. Land that will produce three tons of hay will produce eighteen tons of green forage and a crop of green rye annually, which will give three times the results in dairy products and manure, and that continuously, upon the same land."

It is found to have been demonstrated that the silo governor, invented by Mr. S. M. Colcord, of Dover, Mass., holds the ensilage without heat or fermentation, controlling the operation and changes going on in the silo; that it removes the air, and holds the contents perfectly, precisely as any food is held in air-tight packages; and that it can be applied and used in old as well as new silos.

When we speak of fermentation and heat, the idea is not intended to be conveyed that a silo can be filled with green vegetable matter without eliminating heat. Any succulent vegetable matter piled together in the presence of air commences to heat. If allowed to go on, destructive fermentation sets in, and at length the whole mass becomes putrid and rotten.

ELIMINATING THE AIR.

The various means heretofore used have only measurably arrested this fermentation, and the measure of success has been in just proportion to the exclusion of the air. Not until the invention of Mr. Colcord, a retired druggist of Dover, Mass., have we had means of governing this fermentation at will, or in time to prevent more or less destructive fermentation. The means used by him was the result of scientific study, through his knowledge of chemistry and chemical action.

[From the *Indiana Farmer*, May 21, 1887.]

PRESERVING GREEN FOOD.

SOMETHING NEW AND IMPORTANT IN LIVE STOCK ECONOMY.

Mr. S. M. Colcord, of Dover, Mass., one of the best chemists in the United States, has done the live stock industry a great good in solving the question of preserving green food. During the period of strongest opposition to the ensilage methods, it will be remembered that the *Indiana Farmer* maintained that the success of it was "only a question of skill in construction of silos," and that it was "nonsense to say that we could preserve green fruit by the gallon or more, and could not also exclude the air from the silo," and, further, that "genius and science would satisfactorily solve this matter." And so, while the *Farmer* does not lay special claim to prevision, reasoning from known data, its prediction seems to be fulfilled in the invention of Mr. Colcord. At our request, he has furnished us with cuts and some data to explain his invention. Like the splendid chemist and scientist he is, he seems to have gone about this work with the strong common sense that, to preserve green food, the element usually barring that end was to be eliminated, or overcome. The invention plainly goes directly to the point of excluding the air, which causes over-fermentation and undue action upon the food. Mr. Colcord says that the high temperature theory is a fallacy.

[From the *Farm, Field, and Stockman*, 1888.]

THE COLCORD ENSILAGE EXPERIMENTS.

We received on March 10, from Mr. S. M. Colcord, three days from Dover, Mass., a mail package of Indian corn ensilage, as perfect as when in the fresh state. There was no evidence of heat or fermentation, no acidity or considerable change from the green state, the leaves, stalks, and the grains of corn being quite normal; the odor pleasant, like fresh barley must, when freshly taken from the boilers. Last season, we noticed at length and illustrated Mr. Colcord's system of preserving green fodder in the silo, by means of an apparatus (governor) that perfectly excludes the air. We believe now, as we then stated, that it was a scientifically perfect means of preserving any green forage for winter feeding, in a natural state, including its juices and other normal qualities. The journey of three days, simply wrapped in paper, had not essentially altered its qualities. In fact, city horses ate it and whinnied for more. The preserving of ensilage without heat or fermentation is a long step in advance; and we hope to see this process, as now perfected, largely adopted in the West, especially by dairymen who wish to make the best possible winter milk and its products. To the end that our readers may understand in all its details, we give space this week to a communication from Mr. Colcord, as being important to every person who feeds or proposes to feed ensilage, over the wide area in which the *Farm, Field, and Stockman* is circulated.

The following is Mr. Colcord's communication:—

POSSIBILITIES OF PRESERVING GREEN FORAGE.

"Much of my time the past year has been devoted to building a perfect silo, in which I could try any required experiment, to find out a possible way of preserving green forage by cold pressure, that would fairly or nearly represent canned goods. Knowing your wish for the truth in this direction, I send you the results of my labors and the means I used to prove the facts.

"My experiments cover all the ground from planting to feeding, from construction of silo to the machinery and implements for hândling the corn, as well as the preserved forage, also the connection, arrangement, and convenience of the barn and silo.

"My system differs from anything advanced by others, is opposed to the general mode of producing ensilage, and should not be mixed up or confounded with other methods or manipulations, as the results are unlike. The experiments were intended to show the possibilities in preserving green forage, to find out what can be done in that direction, and the way to do it, as a basis for practical working in farm operations.

"The results show that the principal things to be done for success are to have a tight silo, or pit, drained at the bottom outside; to have the walls perpendicular, smooth and level-faced, with a drip pipe from the centre at the bottom to the outside, terminating with a stop-cock. A governor should be placed at the bottom of the pit, connecting with the drip pipe. When the pit is half to two-thirds filled, a second governor should be put in. When the pit is full, the corn should be trodden down level and covered with 2-inch plank, placed directly upon the corn. It should then be weighted or pressed, to give 2 or more feet of juice from the corn at the bottom. A proper observance of the conditions will produce uniform results with entire success."

WHAT MY NEIGHBORS SAY.

Mr. S. M. COLCORD:

Dear Sir,— I wish to add my testimony and faith in your system of preserving green forage. I became very much interested in your experiments while putting up the staging for building your silo, and witnessing your success in getting a perfect pit, and afterwards noticing your care and attention to every detail while I was running the engine and cutter for cutting and elevating the fodder into the pit; and all the while I was building, or rather extending, the barn over the silo. To me, your plans were a marvelous conception, and your success a wonderful achievement. I have seen nothing in ensilage to compare with green forage preserved with your device. I was rather incredulous until I saw your results. But now that I can see such a mass of vegetable matter preserved without heat or fermentation (very nearly like the vegetables put in tin cans, by heat without fermentation), all the air being taken out and its place occupied by juice pressed from the corn, and this cut down vertically from top to bottom as it is being fed out, leaving a hard face continuously, from which juice can be squeezed out of a handful of it taken from any part of the pit; and, added to this, the fact of there being no heat or fermentation in it, or any odor of ensilage from it, or any waste on the top or around the sides, or at the feeding-troughs, even when the cows are eating an average of 70 pounds daily weighed out to them, and yielding double the results of any other feed rations in milk and manure,— all this to me is marvelous, and I congratulate you upon your success.

Very truly yours,
F. W. SAWIN.

DOVER, February, 1888.

WE, the undersigned, living near Mr. Colcord's farm, having assisted him in harvesting the corn and placing it in the pit, and having seen how the silo was built, how the corn was covered and compressed, and now being able to see the results as to quality and quantity of forage, the milk and the manure, with the very small quantity of hay and grain he is feeding, and the fine appearance of the stock fed on this preserved corn, which is daily increasing in weight, also in the yield of milk,— from our knowledge we are happy to indorse his statements: that the contents of the silo are saturated with juice from the corn from top to bottom; that there is no ensilage odor in the barn or silo; that there is no waste of fodder in the silo or in the cribs. We believe that his system of preserving green forage is the true one, and are happy to state that his experiments are a success, and that any one by following his methods may be sure of success every time, with less than half the expense of feeding in the ordinary way.

Warren Blackman.
Irving Colburn.
Granville Colburn.
James Duffield.
James B. Coughlan.

Dover, Feb. 15, 1888.

Mr. Samuel M. Colcord:

Sir,—Thinking you might like to have it, it is with pleasure that I give you this testimony. Having been employed by S. M. Colcord for six months in 1887 upon his silo and barn, from the construction of the walls to the filling and finishing of the silo and barn, having had full knowledge of his theories and experiments, seen them tested and proved, and having examined his tests and methods, and read what he has written and what has been published in the papers about his system and methods, I feel impelled to add my testimony as to the truth of the statements made. They are not exaggerated statements. I know them to be true; for I have daily taken the temperature in the silo myself, and measured the amount of juice at the bottom. I know that there has been no heating up of the corn in the pit, and no odor of ensilage about the barn or silo. I have seen it taken daily from the pit and weighed out to the cows. It is all of uniform quality, with no waste whatever, and I have noticed that it was all eaten up clean.

From what I have seen, I should say that the published statements are rather short of the whole truth instead of being overstated.

H. B. RYERSON.

MILTON, MASS., May 4, 1889.

Mr. S. M. COLCORD:

Dear Sir,— I have been using your governors, as you are aware, for some years past; and I thought you might like to know what my experience has been with them the past two or three years. I must say to you that I like them just as well as I did at first, and feel sure of good results every time I fill the silo. I am not a chemist, and therefore cannot be expected to prove every point chemically, as you do; but I am not aware of having ever had what you call heat and fermentation in my silo since I commenced to use the governors, and feel so well satisfied with my results, especially when I compare them with those of others that do not use the governors, that I do not take any interest in the science or art of the Fry or any other system, because I feel insured against loss, and faith in the governor keeps me from all anxiety.

I feel to congratulate you upon your success in all your experiments, and the perfection of your system, and feel sure that others will feel as I do about it, whenever they adopt your methods, and come to realize the economy and great value of your discovery. The cost of the governor is a mere trifle compared with its use; and I hope you will live to reap the reward you so richly deserve for all the time and money you have spent the last ten years for the benefit of the farmers.

Wishing you every success, I am
Yours sincerely,
T. E. RUGGLES.

Having used Mr. Colcord's silo governor for some time past with great satisfaction, I cheerfully indorse and coincide with all Mr. Ruggles has written in the above letter.
C. L. COPELAND.

DOVER, May 8, 1889.

Mr. S. M. COLCORD:

Being your nearest neighbor, and quite well acquainted with your silo and preserved green fodder operations, I must congratulate you upon your success in all your experiments, which I confess is a most agreeable surprise to me. I have never seen any ensilage that I would care to feed to my stock, and had lost all faith in the article, if I ever had any; but, when I came to see the green forage you preserve, to taste, smell, and squeeze it, to see the cattle eat it, the milk and manure that come from it, the herd doubled, and you for the first time selling hay, my astonishment turned to admiration; and I don't care who knows it. It is in the mouths of all your neighbors,— I mean the facts and the praise, not the forage. If I had not seen it, assisted in harvesting it, knowing the last poor crop, and the year before seeing two-thirds of your crop blown flat to the ground, I could never have believed that such results could be realized. I hope you will be successful in introducing your system for the benefit of farmers generally, that others may reap a like benefit from the silo governor that you are doing.

Yours very truly,
ALLEN F. SMITH.

WHAT THE BUTCHER SAYS.

Mr. S. M. COLCORD:

Dear Sir,— Having read your statements in the papers about your system of preserving green forage in silos, and being familiar with your farm management of stock and the results of your feeding for milk and beef, I wish to add a few words of testimony in addition to what others say about you. I have watched your operations carefully in building silos, raising and harvesting the crop, and, since you commenced feeding the preserved forage, have seen you double the herd, the milk, and the manure the first year; and now what I want to say is what you have done in the way of beef. In former years, I had nothing to complain of. It was like other people's, and often quite rich in tallow; but, since you have been feeding this preserved forage, I notice a big change. I could not help laughing when I read your account of that big cow that you fed the sour ensilage to, waited an hour and a half until I killed her, and then opened her stomach, and the surprise we both felt when we found no gas in the stomach, none of the bad smell usually found in the stomachs of cows just slaughtered, nothing there but the forage, without any acid or bad smell in it. It was exactly as you stated it; and we found the beef was fat and well mottled. But what I want to say is that the later animals that I have slaughtered from your farm are quite an improvement upon that one. Your last ones shrank but 33 per cent. in weight. They were very fat, and tallowed quite light. The meat was unusually well mottled with fat, and sold for first quality. You know I look over your cattle once or twice a week, and what surprised me was how that forage can make the change. The hair on the cows becomes soft and smooth, the skins become fine, soft, and slippery. The cows are gentle and very quiet. In fact, I go into no barns where I see results equal to yours; and the changes go on so rapidly, after they get into yours, that I hardly know the cows. There is no doubt that your forage and the way you feed is the best thing yet to get the best beef.

J. PEMBER.

MEDFIELD, MASS.

PRIVATE CORRESPONDENCE.

New York, March 12, 1887.

S. M. Colcord, Esq.:

Dear Sir,— I have been very much interested in reading your letter, giving an outline of your work for the coming season, in the way of developing and perfecting the production of Colcord's green fodder.

I feel like encouraging you in this work. It is in the right direction; and, with your present experience and freedom from prejudices, I know of no one so well calculated to meet the difficulties connected with the work, and to overcome them. My judgment is that you will succeed, and with success will come one of the greatest improvements ever made to perfect and cheapen the proper food for milch cows.

I shall be glad to hear from you at any time, and to assist in bringing your improvement into notice, etc. Yours very truly,

Isaac W. White.

40 Wall Street, New York.

CHICAGO, Oct. 31, 1888.

S. M. COLCORD, Dover, Mass.:

My dear Sir,— Your interesting letter of the 29th is here, and I would like to publish it. I like these letters that give exact facts in a definite way, and enter into the full details. But it has occurred to me that it would do the most good to you and the public by reserving it until our number of November 17. We are to issue 100,000 of that number, full count. 20,000 of them, at least, will go to dairy and stock men. We are willing to do whatever we can editorially to make known anything that is really sound, and given for the benefit of our readers; but it seems to us that you ought not to lose this opportunity of presenting some more definite advertisement, and, if possible, some cut of your governor, etc.

Yours very truly,
HOWARD & WILSON PUBLISHING CO.,
C. H. HOWARD, *Editor.*

We note what you say in regard to publishing a book. It is an excellent plan, and you ought to carry it out.

CHICAGO, April 25, 1888.

Mr. S. M. COLCORD:

Dear Sir,— It is our intention to give up the greater part of our issue of June 2 to a thorough discussion of the important subject of ensilage. We should like to have your circulars, and all the information you can give us in regard to this subject, more particularly in regard to your process of preserving sweet ensilage. We received your sample of sweet ensilage some time ago, and, as you are doubtless aware, gave you a very nice notice.

Our object is to educate our readers, and furnish them with the latest and best information on this important subject; and we spare no pains or money in making the article as complete as possible. If we can do anything for you in this issue, we shall be happy to do so. Kindly let us hear from you.

Yours very truly,
JAS. W. WILSON.

CHICAGO, June 21, 1888.

S. M. COLCORD, Esq.:

Dear Sir,—Your favor of June 18 to Mr. Wilson is at hand. We are much interested in your letter. Would you object to our publishing it? It gives new light upon your methods. Of course, we want the paper which you promise as to results. Meanwhile, we must keep alive the interest in your system. Hoping for your permission to publish the letter, we remain very sincerely yours,

              C. H. HOWARD, *Editor in chief.*

              NEW YORK, April 2, 1888.
S. M. COLCORD, Esq.:

Dear Sir,—Your favor of 29th ult. at hand. I feel very much interested in the discussion of the ensilage question. It seems to me that the time is at hand for pushing the discussion, and developing your views as to the proper way of preserving corn and clover for future use. I believe you are the best posted man in the country, both practically and scientifically, and therefore able to educate the farmers as to their interests in this one matter.

    The question clearly is, Will cold storage or hot storage make the best article? The more it is discussed, the better for you as well as the farmers. Unquestionably, cold storage processes are the best, and will produce the best results.

I hope at an early day to see the discussion carried on in the *Country Gentleman*, published in Albany, which has an immense circulation, and will aid greatly in bringing your ways and ideas to the notice of all advanced farmers.

I feel that we are on the eve of a great change of feeling in regard to ensilage. The *Rural New Yorker* is soon to issue a number devoted to a discussion of the question. They have already some points to publish from you. I hope you will keep pushing; and, when you hear of an advanced man in the farming line, send him a circular, write him, and interest him, if possible. Yours truly,

Isaac W. White.

New York, July 24, 1888.

S. M. Colcord, Esq.:

Dear Sir,— Your letter, the *Farm, Field, and Stockman*, and the sample of preserved green forage, came duly to hand. I showed the sample to the editor of the *Rural New Yorker*, and left your letter with him as an assistance in writing an article for his paper, which will soon appear, with cuts of forage before feeding, and also that taken from the stomach of the cow killed by you. I will see that you have a copy of the paper.

The *Rural New Yorker* people are very much interested in your way of preserving green corn fodder, and I think are doing much to bring you prominently before the farmers who are studying up the matter.

I think, with a little patience on your part, you will be the authority on the preservation of green fodder.

Very truly,
Isaac W. White.

Boston, July 29, 1888.

My dear Colcord,— The copy of the *Farm, Field, and Stockman,* which you kindly sent me, was duly received. I have read with much interest every word relating to your experiments, which are well stated, and I think must be convincing to every candid, intelligent person interested in the subject of ensilage. When the full results of your system of ensiloing forage come to be known, there can be no doubt about its taking the place of the empirical, half-way work now in vogue amongst the farmers, on account of its less first cost of the silo, or dumping-hole! But time and experience will cure them, and when the effects of your preserved forage on the beef, milk, and butter become known, as they soon will be known, you will see all other methods abandoned.

It is perfectly clear to my mind that your system is the only true one, for obtaining the best and *most profitable results* from the food fed to animals on the farm, and in time it must prevail.

With my kind regards and best wishes,
                    Yours truly,
                            J. C. C.

                NEW YORK, May 24, 1888.

S. M. COLCORD, Esq.:

Dear Sir,—I am in receipt of your letter of the 21st, and your last contribution in relation to the preserved green fodder question. I have read both with great and renewed interest. I was much gratified to see the treatment you received from the *Rural New Yorker.* Their publication on ensilage gave you a really good notice, and was an indorsement. If you send me another sample, I will show it to them, as well as others I am working with. There is one direction in which I wish you would investigate. I find, with most of those having experience in feeding ensilage to milch cows, that their testimony goes to prove that milk, when first taken from the cows, seems to be, and is, all right; but when they ship it,— and it is often thirty-six hours old before it gets into the hands of the consumer,— then it has something offensive to it, and complaint is made at once. We shipped last year 84,000 gallons; and I wish you would test the milk from the use of your preserved forage, not only once, but daily, for ten days, keeping it until it sours, and having good tasters and smellers to test it daily at different times, to see if any offensive flavor or odor can be discovered.

If the milk will stand this test, as ours does, where we feed on corn meal, bran· middlings, cotton-seed meal, oatmeal, hay, and corn fodder dried, then I think you are lucky, because there then will be no objection to the preserved green forage.

Yours truly,

ISAAC W. WHITE.

ROCKY RIDGE, MD., Jan. 26, 1889.

Mr. S. M. COLCORD:

Dear Sir,— I have been thinking of writing you for quite a while, but neglected doing so. Our ensilage turned out fine. Our corn was good, but weather was wet, and very unfavorable when we commenced to fill. Sometimes too wet to get on the field for several days. So I did not give your governor as fair a test as I wished to do, had I been able to fill quicker; but the difference in waste between the two pits is very marked. In the one without governors, there was about a foot of waste around the sides, while, where the governors were, the waste did not average over three inches at any point. I think this due to the long time we were at filling, caused by continued rains, and our inability to get on to the field; but we are thoroughly convinced of the worthiness of the governors, and, if we live, intend building two more large pits this spring, and will send you an order for governors for both pits.

I take the *Farm, Field, and Stockman,* and read all your articles with much interest.

With best wishes, I am
Very truly yours,
WILLIAM H. BIGGS.

WHAT LARGE DAIRYMEN SAY.

NEW YORK, May 1, 1886.

Mr. S. M. COLCORD:

Dear Sir,— Your interesting letters of 26th inst. are at hand, and read with much pleasure. I feel very much interested in all advances and improvements, and particularly so in the direction you have taken. I have been studying the ensilage question for some time, and am well acquainted, theoretically, with the Fry system, and some time ago decided not to try ensilage, because there was too much doubt and uncertainty in the ability to produce the article wanted,— namely, sweet ensilage. If Colcord's green forage, or fodder, can be produced with certainty, and not be confounded with the article known as ensilage, which has a fearfully black eye, then I want to make and use it.

You can hardly appreciate the strong prejudice against ensilage, and the product of the dairy from its use, in this section. It is so strong that there is no use in trying to convince people that it can be made of proper quality to feed to milch cows. The only way for milk producers this way is to give up the name of silo and ensilage and introduce a new article. I know of no one so well situated as yourself to open the ball by calling attention to "Colcord's Green Forage, or Fodder," as being a better article of food than ensilage, one being a product of fermentation, and the other the preservation of the natural conditions of the fodder, stored for future use.

I see nothing in your past movements to prevent you at once assuming this position, to attract the attention of farmers, and particularly those having silos and using ensilage. The latter can be informed how to make good their expenditures in buildings for trying to make a quality of ensilage fit for dairy purposes.

Please send a circular to Mr. ——, at ——, who has a large silo that has not been used for several years, and is sending milk to this market from about 200 cows.

Yours truly,
I. W. W.

PRESERVED GREEN FORAGE FED TO YOUNG CALVES.

I have been in the habit of raising a few of my best calves. They are considerable care and trouble, before and after they are weaned, and require much attention to give them a good start in life. I had one dropped last year, after my silo was empty. It is now nine months old, and has been fed the past five months on my preserved green forage, with a little shorts, and is looking very nice and healthy. I also have three calves, which I propose to raise. One is five weeks old, one four weeks old, and one two weeks old. The oldest had placed before her, when two weeks old, a box of this forage, with a little ground oats strewn over it, and a bucket of warm water, with a little ground oats stirred in, placed near it; and she began tasting and feeding upon it. In a few days, she ceased to want or take any milk, and when a month old was eating this food, drinking water, and chewing her cud like a cow. The one four weeks old, being in the same pen, began doing the same thing when two weeks old, and now feeds from the same box, contesting with her mate to see which can eat the fastest. The one two weeks old is a large Holstein. Her sire weighed 2,140 pounds. It was very weak when dropped, very inanimate, and would not take a quart of milk a day. Her fæces were very loose and pale yellow, and appeared to be growing worse instead of better. When ten days old, being in the same enclosure with the other two, she arose and went for the green forage. The next day, she was frisky, her fecal discharges became normal, she now eats the same rations with the others, all of them doing well. The oldest of the three is eating too much to make a comely appearance. I shall continue to feed them on this forage and shorts, and watch the results. If it should prove to be successful, and that calves two weeks old will wean themselves on this forage, merely placed before them, that it will regulate their bowels, bring about healthy discharges, keeping them so young in the best condition for growth and improvement, it would seem to be a very important matter to stockmen.

It has been my experience with cows that any disturbance of their natural functions, like failing appetite, scours, or even a gargety condition, is readily counteracted by feeding on this forage alone; but, as corn or any one kind of food is not a perfect ration, the better treatment is to have a small quantity of shorts or wheat bran fed with it, and not feed green forage alone for any great length of time. I don't wish to be misunderstood in these statements. I have had no personal experience in feeding ensilage, and these remarks apply to this preserved green forage only, made without heat or fermentation; but from the examinations I have made, and what I have seen, I should judge that ensilage, as generally made and fed, would have an opposite effect.

[From the *Massachusetts Ploughman*, May 29, 1886.]

THE "SILO GOVERNOR."

In the increasing importance attaching to the silo as an adjunct to the modern system of farming, any improvement proposed for its more perfect operation as an ensilage maker is certain to be welcomed with eager satisfaction. A very great improvement has unquestionably been secured by the combined ingenuity and scientific experience of S. M. Colcord, which was made the subject of investigation at a special meeting of farmers in the Hall of *Ploughman* Building, in which a large number of ensilage experts took part, and a phonographic report of which will be read on the first page of the present issue of the *Ploughman*. It appears that Mr. Colcord's attention was first attracted to the fact that the contents of no two silos or the fermentations in them were ever alike. It was by no means certain that, if a silo made good ensilage one year, it would do so the next. From a series of carefully made investigations into this fact, he felt that he had hit upon the discovery of the real cause and its remedy. Out of these observations and this study was evolved the "Silo Governor." His principal object was to prevent the development of bacteria in the process of putrid fermentation. The effect of the governor on the air in the silo, as the ensilage settles down, is to take it out. In what way it is done, the explanation of Mr. Colcord clearly shows; and he adds the highly important statement that the green forage, under his improvement, comes out without showing any of the usual results of heat and fermentation, and, with no destruction to the corn, guaranteed a superlatively good quality to the product.

In the silo, as generally worked, there is no way for the air to escape except at the top;

and this it cannot do, as a rule, because of the extremely heavy weighting and the close packing. In consequence, the heat and fermentation that take place cause the decomposition of the air, whose component parts form new compounds with other gases generated by the fermentation, thus largely disposing of the air and gas. The bacteria obtain their needed oxygen from the sugar and starch in the corn, and thus detract from the quality of the ensilage according to the amount of destruction caused. The silo governor, it is claimed by its inventor, arrests this work of destruction going on in the green forage; if the air is got out of the ensilage, there will be no heat, and the fodder will be kept in a natural state; if the juice takes the place of air which has been pressed out, it will remain there, and the corn fodder will be preserved in it as in syrup or vegetable extract. Sugar will be found in the heavy butts, and the ensilage will be preserved in better condition for food. Testimonials are offered that, by using the silo governor, the ensilage has less acid, is of a better color, has no odor, and is moister from top to bottom. And the ensilage holds out better, besides, being so much more solid in feeding. The value of the governor is stated in a series of intelligible points elicited in the discussion at the farmers' meeting; and we would advise a careful perusal of our phonographic report of the entire proceedings.

ABOUT FERTILIZERS.

There is no doubt that fertilizers will continue to receive the close attention of farmers. The importance of the subject increases daily, and deserves the careful attention of all cultivators of the soil. The information that has come to us through the laboratory, science, and theory, is very great; yet the information imparted to the farmer has not been put in such shape as to enable him to reap the full benefit, and the farmer has not availed himself of the benefits to himself and his land that might have been derived, had he carefully studied the subject and made careful experiments on his land. But, at the present time, the outlook seems to be more encouraging; and we may safely look for better results in the future, as investigations are now being made that promise to give new and better light in addition to what we now have.

The substances composing fertilizers, their relative quantities and values, are now better known. Those that are supplied from the air, those that are volatile, and those that are inorganic, with formulæ for different crops, are now accessible to farmers who will take the pains to investigate and learn to experiment; and any one can buy the individual raw materials which his land or crops require, and learn to draw largely from nature instead of all from the manufacturer, and can compound them to his mind, or have them made *available* for use and compounded in the required proportions.

Mr. Benjamin Randall, Chelsea Street, East Boston, a thoroughly reliable miller of many years' experience in the business, has done work of this kind for me with satisfactory results. The gain supposed to be made in this way is by bringing the producer and consumer together, saving the profits of the middleman.

Those who cannot designate what they want would do well to try the complete fertilizer made by J. A. Tucker & Co., 13 Doane Street, Boston, known as the Bay State Superphosphates. I have used it for my corn crops, with very good results.

What all farmers should do is to make the most of their manure piles. They should feed for manure as well as milk, which can be done conjointly with increased profit. The manure piles may be used as compost heaps by adding to them such articles as will enhance their value for general purposes, special crops, or the special needs of the land. Such articles as yield potash, phosphoric acid, soda, magnesia, nitrogen, and ammonia, may be added weekly, spreading them over the piles, and saving all the droppings, liquid or solid. This I do partially by keeping the piles under cover out of the way of frost, passing a current of air over them whenever I can get it above the freezing point. This removes a large quantity of water, retaining the valuable part of the liquid manure, and keeps the manure of proper consistence for retaining the gases, using for bedding sand, loam, sawdust, straw, and other dry litter, as absorbents and disinfectants. These things add to the manure in quantity and quality, and should be used also to get the best results mechanically, also saving the manure values, that usually escape as gases, by fixing or compounding them as they rise from the piles. To illustrate, if you spread acid superphosphates (which means bone treated with sulphuric acid in excess) over the piles in good mechanical condition for moderate decomposition, the gases which are evolved will unite with the acid phosphates, and become fixed or non-volatile. This acts also as a disinfectant, purifying the air of the stable. The sulphuric acid, the phosphoric acid, and lime of the bone, being non-volatile, are in the best condition to unite chemically and mechanically with the volatile principles. Each ingredient becomes available for plant food. The whole mass is homogeneous, with loss reduced to the minimum.

By feeding for manure such articles as cotton-seed, flaxseed, or oil-cake, shorts, or wheat-bran, while they go to make up a perfect food for the cow, producing a greater flow and better quality of milk, also improving the cattle in flesh, and giving them a fine, comely appearance, it is also stated that more than half their value goes through the cattle into the manure pile. The chemists and the crops seem also to prove it. The farmers ought also to prove or disprove it in their way by feeding a certain number of cows each with and without these articles, carefully weighing their alvine discharges and planting their experiment plats in all other ways alike, then adding half the grain rations to another portion of the manure that does not contain the grain rations, and again using half the grain rations as fertilizer without any manure, also using a full grain ration without manure.

These grain fertilizers to be well mixed with soil, to prevent burning up the seed, and watered equally, to develop their decomposition. I am aware that these experiments would not be scientific or good in *theory*, but they would pay for the time and trouble to the farmer just the same; but, if in any way he should find that more than half the cost of his grain goes into the manure, and thinks it an economical way of making manure, let him not overdo the thing by converting his herd into fertilizer factories. Many herds are already being injured by this crowding process.

Temperance on the farm will apply to the barn and the crops as well as in the house. The end sought should be the greatest profit with the least loss. This can be done, doubly or even trebly, by using this system of making and feeding preserved green forage, and making the most and best use of the manure.

There is one more point about fertilizers: it is commutation, a mechanical change of state in the ingredients; it is the difference in value of bone between ground or granulated bone and fine or flour of bone. The one is *available* the second or third year after being applied to the land: the other is *available* the first season. I use the term as somewhat analagous to the term used about phosphoric acid, and illustrate it as it may be applied to iron. If you simply pulverize iron very fine, you can so change its condition that you can set fire to it with a match, and burn it up. So you may commute bones so fine that they will become *available* for plant food very soon after being applied to the damp earth, and all plants take their food in liquid form. The reader may think that these statements should be commuted, to make them available, so I will state that, when commercial fertilizers were in their infancy, I bought of this same J. A. Tucker, before referred to, who was then comparatively juvenile and the first superphosphate man I ever heard of. I took some of his fresh ground and fleshy bones, ranging in size from the point of a needle to a thumb-nail, and applied it to a grape-vine in a sickly condition vegetating in a small back yard in the city, and raised the sickly, emaciated vine to the top of a three-story building, training it to the sunshine, where I raised quantities of grapes upon it annually, the phosphates being *immediately* and continuously *available.*

I also visited the works of the same Mr. Benjamin Randall before mentioned, and examined the various mills in use by his predecessors. One of these mills commuted so fine that a lump of anthracite coal the size of a hen's egg would fill a quart bottle, and another would pulverize hard cobble-stones, the size of my two fists, as fast as I could throw them into the mill. Mr. Randall was employed by his predecessors, the mills afterwards coming into his possession, and from long experience is the proper person in the proper place to produce the best ground bone. The proper amount of commutation is a prominent factor in the manufacture of fertilizers.

It would be out of place for me to speak of fertilizers, except in a general way, because in this little book I try to confine myself to statements of facts, and say little about what I don't know; and fertilizers to me are something like oleomargarine and sausages,— vague and uncertain in composition, yet having a value as food for vegetables and an improved food for vegetarians, although unknown compounds, prescribed as simples.

[From the *New England Farmer*.]

SILOS AND ENSILAGE.

THE FAVORABLE REPORTS CONTINUE.— SOME NOTABLE EXPERIMENTS.— AN AIR-TIGHT SILO.

One of our letters of inquiry about ensilage went to Mr. S. M. Colcord, of Dover, Mass., whose response was an invitation to come and see his silo, which a representative of the *Farmer* and *Homes* promptly accepted. Mr. Colcord has had a number of years' experience with ensilage, and believes in it thoroughly; but he also believes that many ensilagists allow the contents of their silos to heat and ferment so as to be seriously injured. He makes the broad assertion that nine-tenths of all the ensilage produced fails to fulfil its possibilities. He is a chemist as well as farmer, and has found in some samples of ensilage acetic acid,— in a cow's daily ration of 60 pounds, an acidity of acid strength equal to three gallons of vinegar of standard strength. Ensilage so sour or which is partly putrid he believes to be unhealthy. Consequently, he has spent much time and money in experiments looking toward a perfect control of the contents of the silo, with a view to making ensilage more similar to canned fruits than the sour, odoriferous product found in many barns. This is to be accomplished by removing the air from the silo while filling and after it is filled, as air is the medium in which the bacteria of fermentation and decay are conveyed from place to place. To prove his theories, he built a new silo in the spring, and constructed it with such thoroughness and accuracy that we give an illustration of the appearance of things when the silo was under way, in order to emphasize the care that was taken to get accurate results. The picture, made from a photograph, shows the framework and the walls of the silo just beginning to rise. The silo is 32 x 12, 20, with the cement walls 17 feet high and 3 feet of plank above this. This plank annex is filled at first; but, after the ensilage settles, none is intended to come above the cement wall. The walls are 18 inches thick, composed of one part cement, two parts coarse sand, two parts small cobblestones, two parts broken stone, mixed with about 30 gallons of water to each barrel of cement. The foundation below the wall is 20 wide and 16 to 20 inches deep, built of cobble-stone and fine gravel, made level on top, and thin cement mortar poured over it and finished level. This foundation is surrounded by a tile drain. The land around it inside and outside was also made quite level and hard, to receive the staging. The timbers on each side the wall were 6 x 6 inches, about 1 foot higher than the wall was to be built, and placed about 5 feet apart. Planks 14 to 16 inches wide, planed on the inside in a planing mill to make them of even thickness, were placed on the inside of these timbers so as to slide up easily, as the cement set. This trough was filled but once daily, which set firmly during the night, keeping the walls level. The timbers on the inside of the pit were erected first. They were straight-grained and not twisted. The planks that tied these timbers together at the top, middle, and bottom were 2 x 8 inches, sawed square at the ends, of equal lengths. They were spiked firmly to the upright timbers and cross-braced, taking the utmost pains to keep the timbers exactly equi-distant and plumb. This work was so thoroughly done that the walls of this pit, when the staging was removed, did not vary 1-8 of an inch in length or breadth, from top to bottom. The outside timbers were put up equi-distant to match the inside ones, and the cross pieces of 1 x 6 fence boards, 6 feet long, nailed to the inside and outside timber, and also to the 4 x 4 outside staging support, as shown in the engraving. As the wall was built up, these cross pieces were sawed away, leaving the inside and outside staging separate, but firm, only tied together at the top.

The engraving also shows 6 1¼-inch iron rods built into the wall from the bottom to 4 feet above the top. When the walls were finished, these rods passed through 8 x 8 cross timbers, and were terminated with double nuts.

The silo was filled with corn-stalks and ears cut in ½-inch lengths. A cover of 2-inch splined planks was then fitted over the top, and the joint between the planks and the wall was covered with rubber packing. The planks were then covered with tarred building paper and several inches of fine sand, making the whole as perfectly air-tight as possible. Then 8 2-inch jack-screws were placed between the cover and the 8 x 8 timbers across the top, and the requisite pressure given to the ensilage, forcing out all of the air, giving a density of 50 pounds to the cubic foot. Mr. Colcord has an ingenious arrangement of pipes about the top and bottom, and through this silo, by which he can make investigations. He has lowered a thermometer into it three times a day, and is confident that no heating has taken place. He can also draw off the juices from the bottom.

The silo was opened a few days before the reporter's visit. The opening was made by removing only two of the end planks, and taking out a narrow strip, exposing a minimum amount of ensilage to the air at a time. The ensilage had been so compacted by the intense pressure when it was first put in that it had to be cut down with a hay-knife, slicing off almost like cheese. It was so solid that the ensilage next to the opening remained in position, there being no trouble from "caving in." The ensilage was the sweetest the *Farmer* man ever saw. There was no perceptible sourness or disagreeable taste to it. The odor of it was hardly noticeable in the barn, and it was very moist. Water could be wrung out from samples taken from the top under the cover.

The pipes about the bottom of the silo, that allow the exit of the air, favor the rapid settling of the ensilage as the silo is filled; and, when it was half full, it settled so fast that 3 inches of juice settled at the bottom.

The corn for the ensilage was raised on a field that previously produced one-half a ton of hay to the acre. Five hundred pounds of Tucker's Bay State Superphosphate per acre were applied with Kemp's manure-spreader, and 500 pounds more were applied in drills. The corn was planted with an Eclipse Corn Planter, which dropped 1 kernel every 3 inches, at the rate of 4 acres per day.

[From the *New England Farmer*.]

SWEET ENSILAGE.

The readers of these columns will remember that last fall we printed a very interesting account of a visit to the silo of Mr. Colcord, of Dover, Mass., and illustrated his silo in process of construction. It was built airtight, with smooth, perpendicular walls, the opposite walls being exactly equal distance from each other, so that under pressure the fodder might descend with the least lateral pressure and the covering come down evenly between the walls. Mr. Colcord's aim is to preserve ensilage without its heating or changing in the least, so that it will be veritably analogous to canned goods. On filling the silo, a slight acetic tendency was noticed, which passed away in a short time, something not unlike the change in the acidity of a Baldwin apple from November to May. Carbonic acid was also noticed; and, as this is heavier than air, it was supposed that this in a measure displaced the air in the silo, thus assisting Mr. Colcord in his desire to make the ensilage perfectly free from air, which must be present if there is any fermentation. One very peculiar thing about the contents of the silo, as it filled, was the large amount of juice which accumulated at the bottom; but this was afterwards absorbed, apparently by capillary attraction, and brought even to the very top. In feeding the ensilage, it was taken out in vertical layers; and at any time large quantities of juice can be squeezed from it, even from portions taken from the top. The result of this experiment was a perfectly sweet, juicy ensilage, without any evidence of putrefaction, and only slight acidity. The cattle eat it readily, there is no waste, and no odor from the forage about the barn or on the hands or clothing of those handling it. The silo was filled with mature corn in full milk, just beginning to glaze and ripen, which would have yielded about 100 bushels of shelled corn, so that by the use of this silo he had no corn to husk, shell, or take to the mill.

The ensilage was fed to 19 head of cattle, 17 being milch cows. 65 pounds were fed daily, with some bran to balance the ration. The milk-flow increased, and in this respect the showing was very satisfactory; but the gain in flesh was even more marked. 16 of the cows gained in four months 2,765 pounds, and 16 gained 1,224 pounds in 30 days. 1 of them gained in weight an average, within a small fraction, of 5 pounds daily during the 30 days, another 4 ¼ pounds daily, another 3 5-6 pounds daily. One large cow was fed on 60 or 70 pounds daily for 90 days, during which time she grew in weight 2 pounds and 5 ounces daily. She was fed 19 pounds of this forage an hour and a half before she was killed; and, after being slaughtered, the contents of her stomach were examined, and found to be sweet, with no offensive odor. The animal was very meaty. Three days before she was butchered, she weighed 1,418.

The 15th of May the silo was three-fourths empty, the corn in the lower part was condensed by pressure to one-half the space occupied by the upper half, but, on account of the absorption of the juice, a cubic foot in the upper half weighed a little more than a cubic foot in the lower half; yet in feeding value they were equal.

All his hay was fed last year to about half the amount of stock, but this year half of the hay will be left over from the use of the silo. The juice from this forage is odorless, agreeable to the taste, and at a temperature of 60 to 80 degrees turns to a pure, weak corn vinegar.

This feeding experiment leads Mr. Colcord to think that the first process of digestion is done in the silo to a great extent, and that a certain quantity of acidity is required in this first process. If it can be done as well or better without using the vital force of the animal, it seems to be a matter of great importance.

ENSILAGE A PROMOTER OF DIGESTION AND ASSIMILATION.

[I regret that I am unable to give the name of the author of the following able article.]

In ensilage there may be a slight loss in the carbo-hydrate elements, and a gain is made in protein, and increased digestibility of the rest, which gives feeding value to what has often been termed the water in ensilage. It is not only easily digested, but also helps digest other richer foods, including grain; and thus, adding the natural juices of plants to the mixed ration, aids nature to assimilate them without calling upon the digestive economy of the animal to do all the work. In the other cases, all this matter is dried down into a hard condition, and must have water to reabsorb it, freshen it up and dissolve it, which requires a good deal of extra force. If you take an apple, you will find the nutriment all in a soluble condition; and, when you take it into the stomach, it is ready to go into the circulation at once. If you dry that apple, all that nutriment becomes like rawhide, and it must be soaked up; and, when you have done that, you have changed its condition. You can never get it back in the same condition it was before the drying was done, and it takes more energy and force to digest that dry food than in its green state. That is the pith of the whole matter.

The nutriment, or the sugar, in dry food is not necessarily changed by the evaporation of the water, but it is simply breaking the chemical union of the water with the rest of the compound; and that chemical reunion has got to be restored by energies of the stomach, which makes extra work and makes it slow. In feeding a cow, you want to give her what she can eat in a given time. A dry feed may contain as much nutriment; but you cannot get as much out of it, because it takes so long to do it that the animal has got to support itself while it is being digested. The point is simply this: that in the green stage the albumen and other matter is, to a large extent, already in solution in a condition in which, when it is separated from the fibrous matter, it can be taken right into the circulation and appropriated. In wetting or steaming fodder, it will help considerably; but it will not overcome the change which the feed undergoes in the desiccation and soaking up again.

[From the *Farm, Field, and Stockman*, March 2, 1889.]

COLCORD'S PRESERVED GREEN FORAGE.

GREAT GAIN IN FLESH.

The following letter has been received from Mr. Colcord: —

Referring to my communication published in the *Farm, Field, and Stockman*, Nov. 17, 1888, after I had filled my silo with very poor green frosted fodder, I now send you the report of my results with that crop.

Preserving Green Forage 133

I pressed the cover down upon the corn slowly, and perfectly level, until I had in the bottom of the silo 20 inches of juice. There was an unusually small quantity of carbonic and acetic acids that came out of the corn, but enough to displace the air in it, and prevent heat and fermentation. Much of these acids was absorbed by the juice, causing a partial vacuum, which, with the pressure, set up capillary attraction, and brought up the juice evenly throughout the mass to the top plank. While this was going on, the temperature was gradually falling in the silo. (I keep a thermometer in the centre of the silo, and examine it daily.) Upon opening the silo to feed from it, we could press out 8 to 10 ounces of the juice from 16 ounces of the forage taken from any part of the silo. Having a surplus quantity of juice at the bottom, we drew it off and fed 6 pounds, mixed with shorts, daily to each cow. This juice was clear, sweet, and nearly odorless.

Twelve of my cows gained in weight, from June to December 1, 1,096 pounds, and were in very good condition; but upon this gain, fed upon this forage 60 pounds daily, with 2 quarts gluten meal and 8 quarts shorts, their gain was 589 pounds additional in thirty days, the milkers nearly doubling their milk, the others gaining in weight.

Considering the quality of the corn put into the silo, I think this is a remarkable showing; and it would seem to be an impossibility for any corn crop to be a failure, with a good silo.

Green forage preserved by this system is better and of greater feeding value during the last month than during the first month, taking it from the silo.

I can see no reason to doubt perfect success every year with the crop, the silo, the system, and the silo governor. By using this system, there is no waste or loss of the corn. By using the governor, all the air and free gases are removed, which prevents heat and fermentation (we have never had above 78° of heat): consequently, we have no foul odors, and 50 per cent. more of this forage can be fed to the cattle than of ordinary ensilage.

THE ADVANTAGE OF JUICE.

There is also a great advantage in having the juice from the corn taken up evenly throughout the mass, reducing the temperature, producing a condition similar to canned goods and superior, inasmuch as the corn is soaking under pressure in free juice, rendering it soft and pulpy, more assimilable as food, and of much greater feeding value.

My stock was fed upon this forage until July 20. After that time, the milk-flow fell off about half. I commenced feeding the preserved forage again, November 25; and during the next thirty-five days the flow of milk doubled from the same cows, and the dry stock increased in flesh, in proportion to the increased flow of milk from the milkers.

As I weigh the fodder every time it is fed, and as the cows are weighed every thirty days, I am able to speak accurately as to results. I have tried to make a comparison with green corn fresh from the field; but, as my cows would not eat a large part of the green stalks, I could not get at a sufficiently accurate estimate for publication. My men who weighed 550 pounds of the fresh green corn, at a feed, thought that amount was about equal to 375 pounds of the preserved green forage.

I put into a very good silo, constructed on my system, last year, two governors for W. H. Bent, Natick, Mass. I have examined the forage that he is now feeding from it to a very fine herd of blooded Holsteins. The quality of the preserved forage was equal to mine, but I have not compared his feeding results with mine. They cannot be the same, because he has taken from one of his cows 40 cans (680 pounds) of milk in ten days. None of my cows will hold as much in twenty-four hours, to say nothing of gaining on that.

I mention this here to show that others can get just as good feed and results by using my system and device as I do. The governors are an economical investment in any silo; but the better the silo, the more perfect will be the preserved fodder.

The importance of this system may be seen when we notice how it differs from ordinary ensilage. By this system, the forage is preserved without heat or fermentation, without foul odor, without any waste; it is continually improving in the silo, is soft and pulpy, and is improving in feeding value while being fed out; as long as it lasts, its quality is improved, its assimilation and feeding value augmented, by soaking in its own juice under pressure, under similar but improved conditions to canned goods.

These conditions are quite the reverse with ensilage. Ensilage is not uniform in quality, and different lots vary very much. Even with the same amount of care, it cannot be depended upon for quality in any case, which accounts for the great number of abandoned silos, although it is often quite good, but it will never bear a good comparison with this preserved forage. It is very seldom that a peck of ensilage can be taken into a warm room and kept a few hours without filling the room with very disagreeable odor; and people who have handled ensilage for a short time, upon entering a warm room, will usually fill it with disagreeable odor, unless their clothing has been previously changed. But this is not the case with this preserved forage. I feed regular rations, weighed, to my cows daily, from 60 to 70 pounds, without any waste, and have fed as high as 85 pounds to large cows. This cannot be done with ensilage. My silo is in my barn; and, even when I am feeding 20 cows in the same barn, people do not notice the odor of ensilage. I am aware that it is difficult for people to understand these statements who have not seen it, but people who have seen these things in my barn attest to these facts.

This system is in operation upon my farm, and is open to the inspection of any one at any time, or to any officers, agents, or Committees of Institutes, Farmers' Clubs, or Granges, wishing to make examinations. As there is no secret about it, all information and every facility is freely given to make examinations; and I will answer any calls upon me to explain this system and device before any meeting of such bodies as Boards of Agriculture or Experiment Stations.

[From the *Farm, Field, and Stockman*, July 14, 1888.]

EXPERIMENTS WITH MILK AND CREAM.

BY S. M. COLCORD.

From a herd of grade cows, fed on Colcord's preserved corn, with half-rations of shorts and cotton-seed, sixteen cows were taken from a rich pasture of fresh grass, and kept in the barn, the temperature being an average of 85°. The increase of milk upon 65 pounds daily ration of this forage was one can daily, without turning the cows out.

June 15, sixteen pints of milk were taken, one pint from each cow's, from the last quart of each milking. It was mixed together, and set by the Cooley system, submerged at a temperature of 48° sixteen hours, the temperature at the end of the setting being 58°. The yield of cream was four and one-half inches to eighteen inches depth of milk, just one-fourth, or 25 per cent. cream.

June 16, sixteen pints of milk taken, one pint each from same cows, fed the same way, setting and temperature the same. This milk was taken from all that the cows gave, the setting was twelve hours, with three inches of cream to eighteen inches depth. Set to twenty-four hours, with about the same amount of cream. The cows not turned out, and held their increase of milk.

June 17, sixteen pints taken, one pint from each cow, from the last half of the milking of each cow, mixed and set the same, temperature of the weather 86°, of the water setting 58°. Yield of cream, four inches to eighteen inches, the cows bellowing in the hot barn for cool water and liberty. The yield of milk fell short one can the past twenty-four hours.

The cream in all these trials was yellow, sweet, and odorless of any taint, and was used upon the table morning, noon, and night, by all the family (eight persons), upon oatmeal, beans, bread, sweet cakes, in tea, coffee, and in several other ways, the quality uniform in all the samples, and the best I have ever seen.

June 17, four days, nine hours, after setting, the milk soured, the cream became acid. Three days, nine hours, after setting, the milk was sweet, the cream slightly acid. Two days, nine hours, after setting, the milk was sweet, the cream sweet. During this time, the weather has varied 30°, with rain, thunder, and lightning. The setting and refrigerator has varied about 10°.

June 20, five days, nine hours, milk and cream both sour. Four days, nine hours, milk acid, cream sour. Three days, nine hours, milk sweet, cream acid.

There has been no odor of ensilage or any bad odor, except that of sour milk or cream. Since the 15th and 16th, the temperature has been kept at 58°. The samples are all uniform, without a taint of foul odor of any kind. The milk and cream are faultless as to color, odor, and taste.

June 26, the curd was partially separated from the whey. The samples are all uniform, without a taint of foul odor of any kind. The taste of curd and whey is very pleasant, cheesy, with no mould, not a suspicion of ensilage taste or odor. The cream is thick, cheesy, of fine odor, and mixed with whey. There is not a shade of taste or smell in the direction of ensilage. These samples are now kept at a temperature of 38°, and are examined frequently. I have never seen samples of milk and cream as pure, sweet, and fine-flavored before. They appear to be faultless. The examinations will continue until further changes are noticed.

I have been induced to make these examinations at the request of parties who furnish the best milk to New York City, for the purpose of knowing whether Colcord's preserved green forage imparts any odor or taste of ensilage to milk or cream. I believe it impossible to get better milk or cream from any other food given to cows.

[From the *Rural New Yorker*.

PATENT SILAGE.

February 1, we received by mail, from Mr. S. M. Colcord, of Dover, Mass., about a pound of silage, which was taken from the silo three days before its arrival. It was the most perfect specimen of preserved fodder we have ever seen, sweet and fragrant. It was sampled by many visitors, several of whom were perfectly willing to put it into their mouths and taste it. We have kept that package of silage on a desk in a warm room ever since. It is now perfectly dry, green and sweet, in far better condition than any corn fodder we have seen.

[Following the above, and in the same paper, there was printed a description of the Colcord process of preserving green forage without heat or fermentation, said description being very similar to that printed on p. 112 of this treatise.]

[From the *New England Farmer,* April, 6, 1889.]

A SUGGESTION FOR THE EXPERIMENT STATIONS.

The *Rural New Yorker*, noticing the ensilage ideas of Mr. S. M. Colcord, which have been noticed several times in these columns, says, "We have long believed that this process of preparing silage will some day revolutionize the ensilage business." The *Rural New Yorker* says in print what the editors of this paper have frequently said to Mr. Colcord personally, that the experiment stations should test this patent system of ensilage-making, side by side with the ordinary silo.

THE PRESERVATION OF ENSILAGE.

BY S. M. COLCORD, DOVER, MASS.

[Published in the Report of the State Board of Agriculture of Pennsylvania, 1888.]

For ten years past, I have made a study of preserving green forage, and a great deal of my time has been spent in visiting silos and making examinations of their contents. There is a great deal of truth in the claims made for ensilage, especially in the direction of assimilation and the digestion of food, analogous in some degree to the difference between grapes and raisins, or between green and dry forage. There is also a great deal of truth in the objections made to ensilage as usually fed to cattle, especially in the direction of its being injured or spoiled by heat and fermentation in the silo, rendering it unfit for wholesome food. It has been found so difficult to preserve it without heat and consequent fermentation that the advice of scientific men has been in the direction of increasing the heat for the purpose of germinating the bacteria in true fermentation and killing them by excess of heat, as they germinate at a temperature above 80° and are killed above 122° Fahrenheit. The objection to this theory and practice is that true fermentation cannot be controlled or stopped before the ensilage becomes unfit for food, as it has been proved that the germs of bacteria will develop after having been exposed to a temperature of 212° for some hours.

I have therefore directed my experiments with a view of preventing heat and fermentation entirely, as the only sure way of preserving green forage; and, in order to ascertain the possibilities of this system, I built a perfectly air-tight silo, with smooth, perpendicular walls, capacity of 6,528 cubic feet, about 150 tons, 384 square feet of top surface, arranged to press with jack-screws, at pleasure, wherever and whenever wanted; filled it with green corn in full milk, cut in half-inch lengths, levelled carefully to level lines, 1 foot apart, striped around the walls with a plumbago pencil; covered it with 2-inch splined plank, and around the sides with 4½-inch rubber packing, then covered with two thicknesses of thick paper, kept in place with 4 inches of sand all over it.

Three governors, consisting of 1-inch iron pipe made into frames 6 by 26 feet, with

¼-inch holes on the under sides, every 6 inches, were placed one at the bottom, one in the centre, and one on the top of this mass of corn, with outlets at the bottom and top, closed with stop-cocks and plugs. These governors were sleeved together, and so arranged that the corn could not stop the ¼-inch holes, but would leave a continuous passage under the pipes for the air and gases to get into the pipes and be conveyed through the pipes to the outside through the openings at top and bottom, thus leaving an air-hole within 3 to 4 feet of every part of the forage. It is not usual to get any juice from the corn in making ensilage, even with 200 pounds weighting to the square foot; but, with this device, the action was so prompt and decided that I had 2 to 3 inches of juice from the corn all over the bottom of the silo before it was three-quarters full, and some days before it was covered. I took the temperature daily at different depths, in the centre, which was never over 72° in any part of the silo. This was about the temperature outside when we commenced filling. Six feet from the bottom, the mercury stood 56° to 58°, which is above the highest temperature we now have, four months after filling. The forage does not heat up when removed from the silo, piled in a heap and left exposed to the air, as ensilage does.

The silo being sealed up air-tight, all the gas coming from it had to come out through the governors at the top of the silo. The next morning after commencing to fill, carbonic acid was abundant. The second day, we had acetic acid, with no rise of temperature. These two acids appeared to be all that came out of the silo. They were very pure and odorless. It is fair to presume that, as carbonic acid is heavier than air, and was present in quantity, it displaced the air in the silo, and that, being readily absorbed by the water or juice, it had a tendency to form a vacuum, which, combined with the pressure and capillary attraction, brought the juice to the very top plank. The silo is now just half empty. The perpendicular face of the cut is $13\frac{1}{2}$ feet. From any part of this face, we can take a handful of the corn and squeeze the juice from it with one hand. The lower half of this mass occupies less than half the space it did when put in. The upper half shows much less pressure; but the weight of a cubic foot of each is about the same, showing the proportion of juice to corn is much greater in the top half. Heavy pressure was kept on for six weeks. During all this time, acetic acid came out pure and pungent.

When that ceased to come, the pressure was discontinued; and, since the pressure has been removed, the acidity has been growing less, a change not unlike the acidity of a Baldwin apple from November to May. My cattle have never objected to the acid. They seem to like it. The good milkers increase in milk, but do not increase so much in flesh; while the young and dry stock increase in flesh, some of them over two pounds daily for sixty days past. I am keeping nearly double the stock of last year, making nearly double the amount of milk and manure, using about the same amount of grain, and employing the same amount of help as last year. I have no weights to put on and take off the silo, no corn to husk, shell, and take to the mill,— it is all in the silo; and I have no corn-fodder to handle, cut up, and steam, and no mangels to cut up and feed. I shall have half my hay left over. It was all fed out last year to about half the amount of stock. I also have enough of this fodder to carry twenty head of stock to the 1st of August.

This system, as developed by experiment and tests, rests mainly upon having tight silos, with smooth, perpendicular, even walls; the opposite walls being at equal distances from each other in all places, so that under pressure the fodder may descend with the least lateral pressure, and the covering may come down evenly without pressure upon the walls; the forage kept spread evenly while being placed in the silo, so that it may be pressed to equal density. The cover should be of 2-inch plank laid across the silo, with 6 x 6 timbers laid across the plank lengthwise of the silo to keep the cover level; and the pressure should be upon the 6 x 6 timber. The best pressure is produced by having 1¼-inch iron rods built into the side walls from the bottom to 6 feet above the walls, placed perpendicular in the centre of the walls, about 4 feet from the end walls and 8 feet distance apart, with 8 x 8 timber (to a 12-foot span) connecting opposite rods, the rods passing through 1½-inch holes through the ends of the timbers, securely fastened on top by double nuts and large, heavy washers under them. The long screws on the rods should be about six threads to an inch. 2-inch jack-screws should be used between the timbers running lengthwise and across the silo.

The heavy strain upon the rods will assist in holding up the walls, the corn can be kept level, and all time, trouble, and expense of weighting avoided.

The corn should be cut fine, ¼ to ¾ inch long. The pressure should give about 30 inches of juice in the bottom of the silo, and is a better guide for pressure than weighting a certain number of pounds, because that amount of juice has been found sufficient under pressure, with the absorption of carbonic acid and capillary attraction to carry the juice to the top of the forage, displacing all air and free gas, representing canned goods by cold pressure instead of heat. This statement must be understood to include the device for removing air and other gases, without which it cannot be done. In large silos, one governor should be placed on the bottom and one in the centre of the silo. These act so promptly that we get juice in the bottom before we get the cover on, and act continuously for six weeks, removing air, carbonic and acetic acid, the forage continually improving in quality from the filling to removing it from the silo. The covering plank is laid directly upon the corn. There is no waste whatever of forage in the silo, or at the feeding-trough. There is no odor from the forage in the silo, about the barn, or from the hands and clothing after handling it. Cattle will eat one ton per month continuously. If the corn is put in mature, the ears upon the stalk, no corn-meal should be fed with it; but cattle will do better with a light ration of shorts and cotton-seed fed with the forage, as corn in its best conditions is not a perfect food, although there is no better food to feed alone.

In comparing different samples of ensilage, it is often difficult to decide which is best; and it is usually found to be of better quality during the first month after opening the silo; but forage preserved by this system is continually improving.

In comparing it with ensilage, we can feed one-third more of it in a given time; it is much more economical,— there is no waste in preserving or feeding it; there are no foul odors about it, and odor is one of the sure tests of quality.

By this system, we expel the air, carbonic and acetic acids, from the silo, pure and simple. In ensilage, these are disposed of by heat and fermentation, through decomposition, forming deleterious compounds with foul odors, increasing as bacteria fermentation is more or less active. Ensilage usually has heat and active bacteria fermentation in it, which causes nearly all the trouble with it; but, when the governor is used in a good silo, heat and fermentation never occur.

It is of great advantage and a great satisfaction to be able to know just the conditions inside the silo at all times, as the amount of juice gives the pressure wanted as well as the acidity. The temperature informs us as to the fermentation, the ability to examine the gases coming from all parts of the silo, at any time, and to know when they cease and how they can be controlled. The device we call the governor, because it is intended not only to show the conditions, but govern them. It has always acted to prevent heat and fermentation, so that we have never had an opportunity to test in stopping and controlling, although designed for that purpose also.

It is also a great advantage to control the pressure at will. The last heavy pressure by the screws gave us acetic acid through the governor, showing no changes beyond, toward decomposition; and it is to be noted that acetic acid came in the second day after commencing to fill.

The quantity of acetic acid in all samples of ensilage has appeared to be one of the great objections to ensilage, and, in judging it, other greater objections have been overlooked. I have found that the quality of the preserved forage does not depend upon its acidity.

To ascertain the effect of good preserved fodder fed in large quantity, I fed to a large cow from 66 to over 70 pounds daily for 90 days, during which time she gained in weight 2 pounds 5 ounces daily. Her last feed, an hour and a half before she was killed, was 19 pounds of bright, odorless forage, fed alone, the acidity of which was equal to the acidity in 11 ounces of commercial acetic acid. The test was made by pressing 10 ounces of juice from 1 pound of the forage, neutralizing with liquor potash, and testing with litmus. The contents of the stomach were immediately examined in the same way. One pound of it was pressed the same as the sample before it was fed, with same results, 10 ounces of juice from 16 ounces. Nothing found in the stomach but this forage. The humidity the same as before being eaten, and tested so nearly neutral that I could not tell whether it leaned to acid or alkali.

The contents of the stomach were not offensive, not more so than the forage before being eaten, except the animal heat in it. Every part of the animal was perfectly healthy. The beef was fat, very meaty and well mottled. This cow, before she was fed on this forage, was in quite ordinary condition, and it was feared that she would not fatten enough for fair beef; but 3 days before she was butchered she weighed 1,418 pounds. How that quantity of acid was disposed of in one and one-half hours by the animal will require more time and further experiments. No animal could have been in better health than she was during the 90 days, or show a more healthy condition of every part upon examination. Her gain in weight for the last 90 days was gradual and continuous, averaging 2 pounds 5 ounces daily. Over 3 pounds of the contents of her stomach is now in the same condition as on the day it was taken out. It has been kept in a tin lard-pail 30 days, is odorless, and seems to be just in the condition for mastication.

This acid forage was taken from the centre of the silo around the perpendicular pipes that very loosely sleeved the governors together, giving the forage an opportunity to absorb a larger portion of the acid passing the loose openings. I do not know what the odor of ensilage would be under like circumstances, after being exposed to heat and fermentation, but presume the foul odors would be very much increased, as is the case with juice from ensilage after exposure to the air.

May 15, 1888.— My silo is now three-quarters empty. This empty part is dry and odorless. The last quarter we are now using is much improved by remaining in the silo. The cover is tight, but there is no pressure or weight upon it. The vertical cut is 13½ feet. The lower half of the corn is condensed to half the space occupied by the upper half, but a cubic foot of the upper half weighs a little more than a cubic foot of the lower half; yet in feeding value they are equal. I feed all the cattle with every ration weighed to them, which is now 65 pounds daily to each animal, reduced from 70 pounds, because the feeding value of the forage is improved by soaking five months in its own juice. The flow of milk is much greater, as well as the gain in flesh. Nothing leaks out or runs down from this vertical cut, and I can now press out 11 ounces of juice from 16 ounces of the forage taken from any part of the silo. 16 of my cows have gained the past 4 months 2,765 pounds, and 16 of them 1,242 pounds during the last 30 days. One of them has gained in weight an average of a fraction over 5 pounds daily during the past 30 days; another one, 4¼ pounds daily; another, 3 5-6 pounds; another, 3 1-3 pounds; another, 3 1-6 pounds; 2 more, 3 pounds daily each, and so on,—all during the last 30 days, with about half my former rations of grain and about one-quarter ration of hay for a change.

The juice from this forage is odorless, agreeable to the taste, and changes but very little upon exposure to the air. It settles clear, and loses much of its acidity. At a temperature of 60° to 80°, it will gradually turn to pure, odorless, weak corn vinegar.

The year previous to building this silo, I fed at the rate of 140 bushels of shelled corn, in the shape of a mixture of cob-meal and oats, to eleven cows, with good results; but I thought I was feeding too much corn for the health of the cattle. This year, I think I am feeding mature corn in full milk that, if allowed to glaze and ripen, would yield about 500 bushels of shelled corn. This has been and is being fed to 19 head of cattle, 17 being milch cows giving 20 cans of milk, 17 pints each, the milk increasing at about the same rate as of flesh, as stated above, in the past 30 days.

This amount of corn goes into the cows as juice, or extract of corn, as between 60 and 70 per cent. is contained in the forage as free juice, held there by absorption; and, fed in this way, I do not consider that it acts in digestion as corn-meal does, and accounts for the greatly increased feeding value of corn forage preserved in this way.

I have not yet tried the experiment of feeding a large quantity of grain with this very moist forage, but I think it can be done to some extent without injury to the cattle. There must be some limit to it, but I have not yet found it. I am not feeding my usual quantity of grain, on account of the enormous increase of flesh, as stated above.

If, as these experiments seem to indicate, the first process of digestion is done in the silo to a great extent, and if it is necessary that a large quantity of acid is required in this first process of digestion, and if we can do it as well or better without using the vital force of the animal, it would seem to be a matter of very great importance.

The cattle eat these rations in half an hour. An hour afterward, the cud is in the best possible condition ready to be chewed over. There is no inflation of gas in the stomach, no acid, no odor. The cows are quiet and docile. The increase of milk, flesh, and manure, is very large. The economy in time, trouble, labor, and expense, is very great, and the results foot up from double to treble any other known methods.

I have not yet made any experiments for the quality of milk, cream, and butter;* but I have no reason to doubt a corresponding advance in both quantity and quality. The quantity we are sure of, and the large amount of hydro-carbons that go to make cream and fat are to be found in corn, bran, cotton and flax seed. Cows fed with this forage in proper quantities (not large) will give very rich and yellow milk, cream, and butter. There is not any objectionable odor in this preserved forage, from the silo to the second stomach of the cow; but it may lack the fine flavor of clover or new-mown hay.

PROGRESS MADE IN PRESERVING GREEN FORAGE IN SILOS.

Men make progress in any direction when they keep their eyes open in all directions, with the mind in an affirmative state, and the will ready to receive the truth and act upon it, when found. I have had good and abundant reasons for pursuing this course in these investigations, and have found occasion to modify and change my opinions in much of the detail of my work, abandon much that I sup-

---

*See p. 138 for recent experiments on milk and cream.

posed was true and reliable, and to hang up as doubtful some things I had considered scientific facts.

Any one attempting to fathom the depths and mysteries of fermentation will find himself in a broad road, with no sharp lines, like wheel-ruts, to guide him, but more like the path of the rainbow, shading and blending, yet never going in a straight line, but always pointing in one direction.

I should never have known what I now do about fermentation, as regards its operations in green forage in silos, had I continued to follow in the direction of other investigators; but when I built a perfect silo, large enough to work the processes in quantity, in which I could try any required experiments, find out all that was going on in the silo, examine all the gases that came from it, ascertain the temperature in it at any time and at any depth, press it level, and enough to get free juice in it from bottom to top, to make the forage very nearly represent canned goods, to prevent heat, fermentation, and foul odor of any kind, and be able to remove the forage in perfect condition, and feed it out without change, in the coldest weather in winter or the warmest weather in summer, I found all my theories and hopes more than realized, because two-thirds of all the difficulties I expected to encounter were removed when all the air was out and my *friend, carbonic acid,* was in. But let no one think that, when he tries an experiment and makes a failure, it has no value. The failures of others have been the landmarks to guide me to success in this matter. Whenever I saw failures in any silo, I was not long in discovering the cause. If it was fungus growth or black rot, I found that it was caused by air getting in from the outside. If I found true fermentation, I always found it produced by air not being removed from the silo, producing heat and fermentation, with decomposition and recombinations, evolving foul odors. I spent a great deal of time and study to find some way of curing these evils, when found in the silo; but, when I became the possessor of a perfect silo, in which I could find the truth of every theory, and prove the facts by actual experiment, my theories and practices became very much modified, and I found two-thirds of my work could be done by avoiding the difficulties, and the practical difficulties so simplified that they could all be met and perfect results obtained, even by persons of ordinary capacity. I learned that by having strong and tight silos, with smooth, level-faced walls, the forage can settle without leaving any cracks or vacant spaces for the air to get in and produce black rot; that removing all the air at the time of filling the silo will prevent heat and fermentation; that pressing out juice from the corn, and bringing it up uniformly throughout the forage to take the place of the air and gases in the forage, will produce a condition like canned goods. This condition can be developed and controlled with very little trouble, and very great economy, by using the silo governor. The details of the processes are all described and explained under appropriate heads, and the system elucidated as well as I know how to do it, in this little book.

While I do not claim that further progress is impossible, I do claim that what we now know is quite sufficient to insure very nearly, if not absolutely, perfect results, and ought to be satisfactory to the most fastidious, making sure of the best results attained as yet by any system devised for cheapening dairy products, and improving them in quality and quantity, whether it be milk, cream, butter, cheese, beef, or even manure; and that whatever progress is made in the future will be made in the direction pointed out in this "System of Preserving Green Forage without Heat or Fermentation."

www.ingramcontent.com/pod-product-compliance
Lightning Source LLC
Chambersburg PA
CBHW030301170426
43202CB00009B/837